T0329655

IMBALANCED
LEARNING

IMBALANCED LEARNING

Foundations, Algorithms, and Applications

Edited by

HAIBO HE
YUNQIAN MA

IEEE
IEEE PRESS

WILEY

ISBN: 9781118074626

10 9 8 7 6 5 4 3 2 1

CONTENTS

PREFACE

Recent developments in science and technology have enabled the growth and availability of data to occur at an explosive rate (the Big Data Challenge). This has created an immense opportunity for knowledge and data engineering/science research to play an essential role in a wide range of applications from daily civilian life to national security and defense. In recent years, the *imbalanced learning* problem has drawn a significant amount of interest from academia, industry, and government as well, ranging from both fundamental research, machine learning and data mining algorithms, to practical applications problems. The key issue with the imbalanced learning problem is the ability of imbalanced data to significantly compromise the performance of most standard learning and mining algorithms. Therefore, when presented with imbalanced datasets, these algorithms fail to properly represent the distributive characteristics of the data and as a result, provide unfavorable accuracies across the classes of the data.

The goal of this book is to provide a timely and critical discussion of the latest research development in this field, including the foundation of imbalanced learning, the state-of-the-art technologies, and the critical application domains. These topics are well balanced between the basic theoretical background, implementation, and practical applications. The targeted readers are from broad groups, such as professional researchers, university faculties, practical engineers, and graduate students, especially those with a background in computer science, engineering, statistics, and mathematics. Moreover, this book can be used as reference or textbook for graduate-level courses in machine learning, data mining, and pattern recognition, to name a few.

We would like to sincerely thank all the contributors of this book for presenting their cutting-edge research in an easily accessible manner, and for putting such discussions into a concrete and coherent context through the entire book. Without their dedicated work and effort, this book would not have been possible! We would also like to thank Mary Hatcher and Stephanie Loh of Wiley-IEEE for their full support, valuable suggestions, and encouragement during the entire development stage of this book.

Haibo He
Kingston, Rhode Island

Yunqian Ma
Golden Valley, Minnesota

January, 2013

CONTRIBUTORS

Josh Attenberg, Etsy, Brooklyn, NY, USA and NYU Stern School of Business, New York, NY, USA

Rukshan Batuwita, National ICT Australia Ltd., Sydney, Australia

Nitesh V. Chawla, Department of Computer Science and Engineering, The University of Notre Dame, Notre Dame, IN, USA

Sheng Chen, Merrill Lynch, Bank of America, New York, NY, USA

Şeyda Ertekin, MIT Sloan School of Management, Massachusetts Institute of Technology, Cambridge, MA, USA

Nathalie Japkowicz, School of Electrical Engineering and Computer Science, University of Ottawa, Ottawa, ON, Canada and Department of Computer Science, Northern Illinois University, Illinois, USA

Haibo He, Department of Electrical, Computer, and Biomedical Engineering, University of Rhode Island, Kingston, RI, USA

T. Ryan Hocns, Department of Computer Science and Engineering, The University of Notre Dame, Notre Dame, IN, USA

Xu-Ying Liu, School of Computer Science and Engineering, Southeast University, Nanjing, China

Vasile Palade, Department of Computer Science, University of Oxford, Oxford, UK

Gary M. Weiss, Department of Computer and Information Sciences, Fordham University, Bronx, NY, USA

Zhi-Hua Zhou, National Key Laboratory for Novel Software Technology, Nanjing University, Nanjing, China

1

INTRODUCTION

HAIBO HE

Department of Electrical, Computer, and Biomedical Engineering, University of Rhode Island, Kingston, RI, USA

Abstract: With the continuous expansion of data availability in many large-scale, complex, and networked systems, it becomes critical to advance raw data from fundamental research on the Big Data challenge to support decision-making processes. Although existing machine-learning and data-mining techniques have shown great success in many real-world applications, *learning from imbalanced data* is a relatively new challenge. This book is dedicated to the state-of-the-art research on imbalanced learning, with a broader discussions on the imbalanced learning foundations, algorithms, databases, assessment metrics, and applications. In this chapter, we provide an introduction to problem formulation, a brief summary of the major categories of imbalanced learning methods, and an overview of the challenges and opportunities in this field. This chapter lays the structural foundation of this book and directs readers to the interesting topics discussed in subsequent chapters.

1.1 PROBLEM FORMULATION

We start with the definition of imbalanced learning in this chapter to lay the foundation for further discussions in the book. Specifically, we define *imbalanced learning* as the learning process for data representation and information extraction with severe data distribution skews to develop effective decision boundaries to support the decision-making process. The learning process could involve supervised learning, unsupervised learning, semi-supervised learning, or a combination

Imbalanced Learning: Foundations, Algorithms, and Applications, First Edition.
Edited by Haibo He and Yunqian Ma.
© 2013 The Institute of Electrical and Electronics Engineers, Inc. Published 2013 by John Wiley & Sons, Inc.

of two or all of them. The task of imbalanced learning could also be applied to regression, classification, or clustering tasks. In this Chapter, we provide a brief introduction to the problem formulation, research methods, and challenges and opportunities in this field. This chapter is based on a recent comprehensive survey and critical review of imbalanced learning as presented in [1], and interested readers could refer to that survey paper for more details regarding imbalanced learning.

Imbalanced learning not only presents significant new challenges to the data research community but also raises many critical questions in real-world data-intensive applications, ranging from civilian applications such as financial and biomedical data analysis to security- and defense-related applications such as surveillance and military data analysis [1]. This increased interest in imbalanced learning is reflected in the recent significantly increased number of publications in this field as well as in the organization of dedicated workshops, conferences, symposiums, and special issues, [2, 3, 4].

To start with a simple example of imbalanced learning, let us consider a popular case study in biomedical data analysis [1]. Consider the "Mammography Data Set," a collection of images acquired from a series of mammography examinations performed on a set of distinct patients [5–7]. For such a dataset, the natural classes that arise are "Positive" or "Negative" for an image representative of a "cancerous" or "healthy" patient, respectively. From experience, one would expect the number of noncancerous patients to exceed greatly the number of cancerous patients; indeed, this dataset contains 10,923 "Negative" (majority class) and 260 "Positive" (minority class) samples. Preferably, we require a classifier that provides a balanced degree of predictive accuracy for both the minority and majority classes on the dataset. However, in many standard learning algorithms, we find that classifiers tend to provide a severely imbalanced degree of accuracy, with the majority class having close to 100% accuracy and the minority class having accuracies of 0 ~ 10%; see for instance, [5, 7]. Suppose a classifier achieves 5% accuracy on the minority class of the mammography dataset. Analytically, this would suggest that 247 minority samples are misclassified as majority samples (i.e., 247 cancerous patients are diagnosed as noncancerous). In the medical industry, the ramifications of such a consequence can be overwhelmingly costly, more so than classifying a noncancerous patient as cancerous [8]. Furthermore, this also suggests that the conventional evaluation practice of using singular assessment criteria, such as the overall accuracy or error rate, does not provide adequate information in the case of imbalanced learning. In an extreme case, if a given dataset includes 1% of minority class examples and 99% of majority class examples, a naive approach of classifying every example to be a majority class example would provide an accuracy of 99%. Taken at face value, 99% accuracy across the entire dataset appears superb; however, by the same token, this description fails to reflect the fact that none of the minority examples are identified, when in many situations, those minority examples are of much more interest. This clearly demonstrates the need to revisit the assessment metrics for imbalanced learning, which is discussed in Chapter 8.

1.2 STATE-OF-THE-ART RESEARCH

Given the new challenges facing imbalanced learning, extensive efforts and significant progress have been made in the community to tackle this problem. In this section, we provide a brief summary of the major category of approaches for imbalanced learning. Our goal is just to highlight some of the major research methodologies while directing the readers to different chapters in this book for the latest research development in each category of approach. Furthermore, a comprehensive summary and critical review of various types of imbalanced learning techniques can also be found in a recent survey [1].

1.2.1 Sampling Methods

Sampling methods seem to be the dominate type of approach in the community as they tackle imbalanced learning in a straightforward manner. In general, the use of sampling methods in imbalanced learning consists of the modification of an imbalanced dataset by some mechanism in order to provide a balanced distribution. Representative work in this area includes random oversampling [9], random undersampling [10], synthetic sampling with data generation [5, 11–13], cluster-based sampling methods [14], and integration of sampling and boosting [6, 15, 16].

The key aspect of sampling methods is the mechanism used to sample the original dataset. Under different assumptions and with different objective considerations, various approaches have been proposed. For instance, the mechanism of *random oversampling* follows naturally from its description by replicating a randomly selected set of examples from the minority class. On the basis of such simple sampling techniques, many informed sampling methods have been proposed, such as the *EasyEnsemble* and *BalanceCascade* algorithms [17]. Synthetic sampling with data generation techniques has also attracted much attention. For example, the synthetic minority oversampling technique (SMOTE) algorithm creates artificial data based on the feature space similarities between existing minority examples [5]. Adaptive sampling methods have also been proposed, such as the borderline-SMOTE [11] and adaptive synthetic (ADASYN) sampling [12] algorithms. Sampling strategies have also been integrated with ensemble learning techniques by the community, such as in SMOTEBoost [15], RAMOBoost [18], and DataBoost-IM [6]. Data-cleaning techniques, such as Tomek links [19], have been effectively applied to remove the overlapping that is introduced from sampling methods for imbalanced learning. Some representative work in this area includes the one-side selection (OSS) method [13] and the neighborhood cleaning rule (NCL) [20].

1.2.2 Cost-Sensitive Methods

Cost-sensitive learning methods target the problem of imbalanced learning by using different cost matrices that describe the costs for misclassifying any particular data example [21, 22]. Research in the past indicates that there is a strong connection between cost-sensitive learning and imbalanced learning [4, 23, 24]. In general, there are three categories of approaches to implement cost-sensitive learning for imbalanced data. The first class of techniques applies misclassification costs to the dataset as a form of dataspace weighting (*translation theorem* [25]); these techniques are essentially cost-sensitive bootstrap sampling approaches where misclassification costs are used to select the best training distribution. The second class applies cost-minimizing techniques to the combination schemes of ensemble methods (*Metacost framework* [26]); this class consists of various meta techniques, such as the AdaC1, AdaC2, and AdaC3 methods [27] and AdaCost [28]. The third class of techniques incorporates cost-sensitive functions or features directly into classification paradigms to essentially "fit" the cost-sensitive framework into these classifiers, such as the cost-sensitive decision trees [21, 24], cost-sensitive neural networks [29, 30], cost-sensitive Bayesian classifiers [31, 32], and cost-sensitive support vector machines (SVMs) [33–35].

1.2.3 Kernel-Based Learning Methods

There have been many studies that integrate kernel-based learning methods with general sampling and ensemble techniques for imbalanced learning. Some examples include the SMOTE with different costs (SDC) method [36] and the ensembles of over/undersampled SVMs [37, 38]. For example, the SDC algorithm uses different error costs [36] for different classes to bias the SVM to guarantee a more well-defined boundary. The granular support vector machines—repetitive undersampling (GSVM-RU) algorithm was proposed in [39] to integrate SVM learning with undersampling methods. Another major category of kernel-based learning research efforts focuses more concretely on the mechanisms of the SVM itself; this group of methods are often called *kernel modification methods*, such as the kernel classifier construction algorithm proposed in [40]. Other examples of kernel modification include the various techniques used for adjusting the SVM class boundary [41, 42]. Furthermore, the total margin-based adaptive fuzzy SVM (TAF-SVM) kernel method was proposed in [43] to improve SVM robustness. Other major kernel modification methods include the k-category proximal SVM (PSVM) [44], SVMs for extreme imbalanced datasets [45], support cluster machines (SCMs) [46], kernel neural gas (KNG) algorithm [47], hybrid kernel machine ensemble (HKME) algorithm [48], and the Adaboost relevance vector machine (RVM) [49].

1.2.4 Active Learning Methods

Active learning methods have also been proposed for imbalanced learning in the literature [50–53]. For instance, Ertekin et al. [51, 52] proposed an efficient SVM-based active learning method that queries a small pool of data at each iterative step of active learning instead of querying the entire dataset. Active learning integrations with sampling techniques have also been proposed. For instance, Zhu and Hovy [54] analyzed the effect of undersampling and oversampling techniques with active learning for the word sense disambiguation (WSD) imbalanced learning problem. Another active learning sampling method is the simple active learning heuristic (SALH) approach proposed in [55]. The main aim of this method is to provide a generic model for the evolution of genetic programming (GP) classifiers by integrating the stochastic subsampling method and a modified Wilcoxon–Mann–Whitney (WMW) cost function [55]. Major advantages of the SALH method include the ability to actively bias the data distribution for learning, the existence of a robust cost function, and the improvement of the computational cost related to the fitness evaluation.

1.2.5 One-Class Learning Methods

The one-class learning or novelty detection method has also attracted much attention in the community for imbalanced learning [4]. Generally speaking, this category of approaches aims to recognize instances of a concept by using mainly, or only, a single class of examples (i.e., recognition-based methodology) rather than differentiating between instances of both positive and negative classes as in the conventional learning approaches (i.e., discrimination-based inductive methodology). Representative work in this area includes the one-class SVMs [56, 57] and the autoassociator (or autoencoder) method [58–60]. For instance, in [59], a comparison between different sampling methods and the one-class autoassociator method was presented. The novelty detection approach based on redundancy compression and nonredundancy differentiation techniques was investigated in [60]. Lee and Cho [61] suggested that novelty detection methods are particularly useful for extremely imbalanced datasets, whereas regular discrimination-based inductive classifiers are suitable for relatively moderate imbalanced datasets.

Although the current efforts in the community are focused on two-class imbalanced problems, *multi-class imbalanced learning* problems also exist and have been investigated in numerous works. For instance, in [62], a cost-sensitive boosting algorithm AdaC2.M1 was proposed to tackle the class imbalance problem with multiple classes. In [63], an iterative method for multi-class cost-sensitive learning was proposed. Other works of multi-class imbalanced learning include the min–max modular network [64] and the rescaling approach for multi-class cost-sensitive neural networks [65], to name a few.

Our discussions in this section by no means provide a full coverage of the complete set of methods to tackle the imbalanced learning problem, given the variety of assumptions for the imbalanced data and different learning objectives of different applications. Interested readers can refer to [1] for a recent survey of the imbalanced learning methods. The latest research development on this topic can be found in the following chapters.

1.3   LOOKING AHEAD: CHALLENGES AND OPPORTUNITIES

Given the increased complexity of data in many of the current real-world applications, imbalanced learning presents many new challenges as well as opportunities to the community. Here, we highlight a few of those to hopefully provide some suggestions for long-term research in this field.

1.3.1   Advancement of the Foundations and Principles of Imbalanced Learning

Although there are numerous new efforts in the community targeting imbalanced learning, many of the current research methodologies are very heuristic and ad hoc based. There is a lack of theoretical foundation and principles to guide the development of systematic imbalanced learning approaches. For instance, although almost every paper presented to the community claims that there is a certain degree of improvement on learning performance or efficiency, there are situations where learning from the original imbalanced data could provide better performance. This raises important questions: *what is the assurance that algorithms specifically designed for imbalanced learning could really help, and how and why?* [1]. Can one simply design robust-enough algorithms that could learn from whatever data are presented [53]? Also, as there are many existing base learning algorithms such as the decision tree, neural network, and SVM, is there a way we could develop a theoretical guidance on which base learning algorithm is most appropriate for a particular type of imbalanced data? Are there any error bounds for those base learning algorithms for imbalanced data? What is the relationship between data-imbalanced ratio and learning model complexity? What are the best levels of balanced ratio for a given base learning algorithm? All of these are open questions to the community now. In fact, a thorough understanding of these questions will not only provide fundamental insights into the imbalanced learning problem but also provide critical technical tools and solutions to many practical real imbalanced learning applications. Therefore, it is essential for the community to investigate all, or at least some, of these questions for the long-term sustainable development of this field. In Chapter 2, a dedicated section discusses the foundations of imbalanced learning to provide some new insights along this direction.

1.3.2   Unified Data Benchmark for Imbalanced Learning

It is well known that *data* plays a key role in any kind of machine-learning and data-mining research. This is especially the case for the relatively new field of imbalanced learning. Although there are currently many publicly available benchmarks for assessing the effectiveness of different learning algorithm/tools (e.g., University of California, Irvine (UCI) data repository [66], and National

Institute of Standards and Technology (NIST) Scientific and Technical Databases [67]), there are very few data benchmarks that are solely dedicated to imbalanced learning problems. This has caused data for imbalanced learning to be very costly in the society. For instance, many of the existing data benchmarks require additional manipulation before they can be applied to imbalanced learning scenarios for each algorithm. This limitation has created a bottleneck in the long-term development of research in this field. Therefore, unified data benchmarks for imbalanced learning are important to provide an open-access source for the community not only to promote data sharing but also to provide a common platform to ensure a fair comparative study among different methods.

1.3.3 Standardized Assessment Metrics

As discussed in Section 1.1, traditional assessment techniques may not be able to provide a fair and comprehensive evaluation of the imbalanced learning algorithms. In particular, it is widely agreed that a singular evaluation metric, such as overall classification error rate, is not sufficient when handling imbalanced learning problems. As suggested in [1], it seems that a combination of singular-based metrics (e.g., precision, recall, F-measure, and G-mean) together with curve-based assessment metrics [e.g., receiver operating characteristic (ROC) curve, precision–recall (PR) curve, and cost curve) will provide a more complete assessment of imbalanced learning. Therefore, it is necessary for the community to establish—as a standard—the practice of using such assessment approaches to provide more insights into the advantages and limitations of different types of imbalanced learning methods. More details on this can be found in Chapter 8.

1.3.4 Emerging Applications with Imbalanced Learning

Imbalanced learning has presented itself to be an essential part in many critical real-world applications. For instance, in the aforementioned biomedical diagnosis situation, an effective learning approach that could handle the imbalanced data is key to supporting the medical decision-making process. Similar scenarios have appeared in many other mission-critical tasks, such as security (e.g., abnormal behavior recognition), defense (e.g., military data analysis), and financial industry (e.g., outlier detection). This book also presents a few examples of such critical applications to demonstrate the importance of imbalanced learning.

1.4 ACKNOWLEDGMENTS

This work was supported in part by the National Science Foundation (NSF) under grant ECCS 1053717 and Army Research Office (ARO) under Grant W911NF-12-1-0378.

REFERENCES

1. H. He and E. A. Garcia, "Learning from imbalanced data sets," *IEEE Transactions on Knowledge and Data Engineering*, vol. 21, no. 9, pp. 1263–1284, 2009.
2. N. Japkowicz, (ed.), "Learning from imbalanced data sets," American Association for Artificial Intelligence (AAAI) Workshop Technical Report WS-00-05, 2000.
3. N. V. Chawla, N. Japkowicz, and A. Kolcz, (eds.), Workshop on learning from imbalanced data sets II, in *Proceedings of International Conference on Machine Learning*, 2003.
4. N. V. Chawla, N. Japkowicz, and A. Kolcz, "Editorial: Special issue on learning from imbalanced data sets," *ACM SIGKDD Explorations Newsletter*, vol. 6, no. 1, pp. 1–6, 2004.
5. N. V. Chawla, K. W. Bowyer, L. O. Hall, and W. P. Kegelmeyer, "SMOTE: Synthetic minority over-sampling technique," *Journal of Artificial Intelligence Research*, vol. 16, pp. 321–357, 2002.
6. H. Guo and H. L. Viktor, "Learning from imbalanced data sets with boosting and data generation: The DataBoost-IM approach," *ACM SIGKDD Explorations Newsletter*, vol. 6, no. 1, pp. 30–39, 2004.
7. K. Woods, C. Doss, K. Bowyer, J. Solka, C. Priebe, and W. Kegelmeyer, "Comparative evaluation of pattern recognition techniques for detection of microcalcifications in mammography," *International Journal of Pattern Recognition and Artificial Intelligence*, vol. 7, no. 6, pp. 1417–1436, 1993.
8. R. B. Rao, S. Krishnan, and R. S. Niculescu, "Data mining for improved cardiac care," *ACM SIGKDD Explorations Newsletter*, vol. 8, no. 1, pp. 3–10, 2006.
9. A. Estabrooks, T. Jo, and N. Japkowicz, "A multiple resampling method for learning from imbalanced data sets," *Computational Intelligence*, vol. 20, no. 1, 18–36, 2004.
10. C. Drummond and R. C. Holte, "C4.5, class imbalance, and cost sensitivity: Why under-sampling beats over-sampling," in *Proceedings of International Conference Machine Learning, Workshop on Learning from Imbalanced Data Sets II*, 2003.
11. H. Han, W. Y. Wang, and B. H. Mao, "Borderline-SMOTE: A new over-sampling method in imbalanced data sets learning," in *Proceedings of International Conference on Intelligent Computing* (Hefei, China), Springer, pp. 878–887, 2005.
12. H. He, Y. Bai, E. A. Garcia, and S. Li, "ADASYN: Adaptive synthetic sampling approach for imbalanced learning," in *Proceedings of International Joint Conference on Neural Networks* (Hong Kong, China), IEEE, pp. 1322–1328, 2008.
13. M. Kubat and S. Matwin, "Addressing the curse of imbalanced training sets: One-sided selection," in *Proceedings of International Conference on Machine Learning*, pp. 179–186, 1997.
14. T. Jo and N. Japkowicz, "Class imbalances versus small disjuncts," *ACM SIGKDD Explorations Newsletter*, vol. 6, no. 1, pp. 40–49, 2004.
15. N. V. Chawla, A. Lazarevic, L. O. Hall, and K. W. Bowyer, "SMOTEBoost: Improving prediction of the minority class in boosting," in *Proceedings of Principles on Knowledge Discovery Databases*, pp. 107–119, 2003.
16. D. Mease, A. J. Wyner, and A. Buja, "Boosted classification trees and class probability/quantile estimation," *Journal of Machine Learning Research*, vol. 8, pp. 409–439, 2007.

17. X. Y. Liu, J. Wu, and Z. H. Zhou, "Exploratory under-sampling for class-imbalance learning," in *Proceedings of International Conference on Data Mining* (Washington, DC, USA), pp. 965–969, IEEE Computer Society, 2006.

18. S. Chen, H. He, and E. A. Garcia, "RAMOBoost: Ranked minority over-sampling in boosting," *IEEE Transactions on Neural Networks*, vol. 21, no. 10, pp. 1624–1642, 2010.

19. I. Tomek, "Two modifications of CNN," *IEEE Transactions on System, Man, and Cybernetics*, vol. 6, pp. 769–772, 1976.

20. J. Laurikkala, "Improving identification of difficult small classes by balancing class distribution," in *Proceedings of Conference AI in Medicine in Europe: Artificial Intelligence Medicine*, pp. 63–66, 2001.

21. C. Elkan, "The foundations of cost-sensitive learning," in *Proceedings of International Joint Conference on Artificial Intelligence*, pp. 973–978, 2001.

22. K. M. Ting, "An instance-weighting method to induce cost-sensitive trees," *IEEE Transactions on Knowledge and Data Engineering*, vol. 14, no. 3, pp. 659–665, 2002.

23. G. M. Weiss, "Mining with rarity: A unifying framework," *ACM SIGKDD Explorations Newsletter*, vol. 6, no. 1, pp. 7–19, 2004.

24. M. A. Maloof, "Learning when data sets are imbalanced and when costs are unequal and unknown," in *Proceedings of International Conference on Machine Learning, Workshop on Learning from Imbalanced Data Sets II*, 2003.

25. B. Zadrozny, J. Langford, and N. Abe, "Cost-sensitive learning by cost-proportionate example weighting," in *Proceedings of International Conference on Data Mining*, pp. 435–442, 2003.

26. P. Domingos, "MetaCost: A general method for making classifiers cost-sensitive," in *Proceedings of International Conference on Knowledge Discovery and Data Mining*, pp. 155–164, 1999.

27. Y. Sun, M. S. Kamel, A. K. C. Wong, and Y. Wang, "Cost-sensitive boosting for classification of imbalanced data," *Pattern Recognition*, vol. 40, no. 12, pp. 3358–3378, 2007.

28. W. Fan, S. J. Stolfo, J. Zhang, and P. K. Chan, "AdaCost: Misclassification cost-sensitive boosting," in *Proceedings of International Conference on Machine Learning*, pp. 97–105, 1999.

29. M. Z. Kukar and I. Kononenko, "Cost-sensitive learning with neural networks," in *Proceedings of European Conference on Artificial Intelligence*, pp. 445–449, 1998.

30. X. Y. Liu and Z. H. Zhou, "Training cost-sensitive neural networks with methods addressing the class imbalance problem," *IEEE Transactions on Knowledge and Data Engineering*, vol. 18, no. 1, pp. 63–77, 2006.

31. P. Domingos and M. Pazzani, "Beyond independence: Conditions for the optimality of the simple Bayesian classifier," *Machine Learning*, 105–112, 1996.

32. G. R. I. Webb and M. J. Pazzani, "Adjusted probability naive Bayesian induction," in *Australian Joint Conference on Artificial Intelligence* (Brisbane, Australia), Springer, pp. 285–295, 1998.

33. G. Fumera and F. Roli, "Support vector machines with embedded reject option," in *Proceedings of International Workshop on Pattern Recognition with Support Vector Machines*, (Niagara Falls, Canada), pp. 68–82, Springer-Verlag, 2002.

34. J. C. Platt, "Fast training of support vector machines using sequential minimal optimization," Bernhard Schölkopf, Christopher J. C. Burges, Alexander J. Smola, (Eds.), in *Advances in Kernel Methods: Support Vector Learning*, pp. 185–208, Cambridge, MA: MIT Press, 1999.

35. J. T. Kwok, "Moderating the outputs of support vector machine classifiers," *IEEE Transactions on Neural Networks*, vol. 10, no. 5, pp. 1018–1031, 1999.

36. R. Akbani, S. Kwek, and N. Japkowicz, "Applying support vector machines to imbalanced datasets," *Lecture Notes in Computer Science*, vol. 3201, pp. 39–50, 2004.

37. Y. Liu, A. An, and X. Huang, "Boosting prediction accuracy on imbalanced datasets with SVM ensembles," *Lecture Notes in Artificial Intelligence*, vol. 3918, pp. 107–118, 2006.

38. B. X. Wang and N. Japkowicz, "Boosting support vector machines for imbalanced data sets," *Lecture Notes in Artificial Intelligence*, vol. 4994, pp. 38–47, 2008.

39. Y. Tang and Y. Q. Zhang, "Granular SVM with repetitive undersampling for highly imbalanced protein homology prediction," in *Proceedings of International Conference on Granular Computing*, pp. 457–460, 2006.

40. X. Hong, S. Chen, and C. J. Harris, "A kernel-based two-class classifier for imbalanced datasets," *IEEE Transactions on Neural Networks*, vol. 18, no. 1, pp. 28–41, 2007.

41. G. Wu and E. Y. Chang, "Aligning boundary in kernel space for learning imbalanced dataset," in *Proceedings of International Conference on Data Mining*, pp. 265–272, 2004.

42. G. Wu and E. Y. Chang, "KBA: Kernel boundary alignment considering imbalanced data distribution," *IEEE Transactions on Knowledge and Data Engineering*, vol. 17, no. 6, pp. 786–795, 2005.

43. Y. H. Liu and Y. T. Chen, "Face recognition using total margin-based adaptive fuzzy support vector machines," *IEEE Transactions on Neural Networks*, vol. 18, no. 1, pp. 178–192, 2007.

44. G. Fung and O. L. Mangasarian, "Multicategory proximal support vector machine classifiers," *Machine Learning*, vol. 59, no. 1, 2, pp. 77–97, 2005.

45. B. Raskutti and A. Kowalczyk, "Extreme re-balancing for SVMs: A case study," *ACM SIGKDD Explorations Newsletter*, vol. 6, no. 1, pp. 60–69, 2004.

46. J. Yuan, J. Li, and B. Zhang, "Learning concepts from large scale imbalanced data sets using support cluster machines," in *Proceedings of International Conference on Multimedia*, (Santa Barbara, CA, USA), pp. 441–450, ACM, 2006.

47. A. K. Qin and P. N. Suganthan, "Kernel neural gas algorithms with application to cluster analysis," in *Proceedings of International Conference on Pattern Recognition*, IEEE Computer Society, (Washington, DC, USA), 2004.

48. P. Li, K. L. Chan, and W. Fang, "Hybrid kernel machine ensemble for imbalanced data sets," in *Proceedings of International Conference on Pattern Recognition* (Hong Kong, China), pp. 1108–1111, 2006.

49. A. Tashk, R. Bayesteh, and K. Faez, "Boosted Bayesian kernel classifier method for face detection," in *Proceedings of International Conference on Natural Computation*, pp. 533–537, IEEE Computer Society, (Washington, DC, USA), 2007.

50. N. Abe, Invited talk: "Sampling approaches to learning from imbalanced datasets: Active learning, cost sensitive learning and beyond," in *Proceedings of International Conference on Machine Learning, Workshop on Learning from Imbalanced Data Sets II*, 2003.

51. S. Ertekin, J. Huang, L. Bottou, and L. Giles, "Learning on the border: Active learning in imbalanced data classification," in *Proceedings of ACM Conference on Information and Knowledge Management*, pp. 127–136, ACM, (Lisboa, Portugal), 2007.

52. S. Ertekin, J. Huang, and C. L. Giles, "Active learning for class imbalance problem," in *Proceedings of International SIGIR Conference on Research and Development in Information Retrieval*, pp. 823–824, 2007.

53. F. Provost, "Machine learning from imbalanced datasets 101," Learning from Imbalanced Data Sets: Papers from the American Association for Artificial Intelligence Workshop, Technical Report WS-00-05, 2000.

54. J. Zhu and E. Hovy, "Active learning for word sense disambiguation with methods for addressing the class imbalance problem," in *Proceedings of Joint Conference on Empirical Methods in Natural Language Processing and Computational Natural Language Learning* (Prague, Czech Republic), pp. 783–790, Association for Computational Linguistics, 2007.

55. J. Doucette and M. I. Heywood, "GP classification under imbalanced data sets: Active sub-sampling and AUC approximation," *Lecture Notes in Computer Science*, vol. 4971, pp. 266–277, 2008.

56. B. Scholkopt, J. C. Platt, J. Shawe-Taylor, A. J. Smola, and R. C. Williamson, "Estimating the support of a high-dimensional distribution," *Neural Computation*, vol. 13, pp. 1443–1471, 2001.

57. L. M. Manevitz and M. Yousef, "One-class SVMs for document classification," *Journal of Machine Learning Research*, vol. 2, pp. 139–154, 2001.

58. N. Japkowicz, "Supervised versus unsupervised binary-learning by feedforward neural networks," *Machine Learning*, vol. 42, pp. 97–122, 2001.

59. N. Japkowicz, "Learning from imbalanced data sets: A comparison of various strategies," Proceedings of Learning from Imbalanced Data Sets, the AAAI Workshop, Technical Report WS-00-05, pp. 10–15, 2000.

60. N. Japkowicz, C. Myers, and M. Gluck, "A novelty detection approach to classification," in *Proceedings of Joint Conference on Artificial Intelligence*, pp. 518–523, 1995.

61. H. J. Lee and S. Cho, "The novelty detection approach for difference degrees of class imbalance," *Lecture Notes in Computer Science*, vol. 4233, pp. 21–30, 2006.

62. Y. Sun, M. S. Kamel, and Y. Wang, "Boosting for learning multiple classes with imbalanced class distribution," in *Proceedings of International Conference on Data Mining*, pp. 592–602, IEEE Computer Society, (Washington, DC, USA), 2006.

63. N. Abe, B. Zadrozny, and J. Langford, "An iterative method for multi-class cost-sensitive learning," in *Proceedings of the Tenth ACM SIGKDD International Conference on Knowledge Discovery and Data Mining* (Seattle, WA, USA), pp. 3–11, ACM, 2004.

64. K. Chen, B. L. Lu, and J. Kwok, "Efficient classification of multi-label and imbalanced data using min-max modular classifiers," in *Proceedings of World Congress on Computation Intelligence - International Joint Conference on Neural Networks*, pp. 1770–1775, 2006.

65. Z. H. Zhou and X. Y. Liu, "On multi-class cost-sensitive learning," in *Proceedings of National Conference on Artificial Intelligence*, pp. 567–572, 2006.

66. UC Irvine Machine Learning Repository [online], available: http://archive.ics.uci. edu/ml/ (accessed January 28, 2013).

67. NIST Scientific and Technical Databases [online], available: http://nist.gov/srd/ online.htm (accessed January 28, 2013).

2

FOUNDATIONS OF IMBALANCED LEARNING

GARY M. WEISS

Department of Computer and Information Sciences, Fordham, University, Bronx, NY, USA

Abstract: Many important learning problems, from a wide variety of domains, involve learning from imbalanced data. Because this learning task is quite challenging, there has been a tremendous amount of research on this topic over the past 15 years. However, much of this research has focused on methods for dealing with imbalanced data, without discussing exactly how or why such methods work—or what underlying issues they address. This is a significant oversight, which this chapter helps to address. This chapter begins by describing what is meant by imbalanced data, and by showing the effects of such data on learning. It then describes the fundamental learning issues that arise when learning from imbalanced data, and categorizes these issues as problem-definition-level issues, data-level issues, or algorithm-level issues. The chapter then describes the methods for addressing these issues and organizes these methods using the same three categories. As one example, the data-level issue of "absolute rarity" (i.e., not having sufficient numbers of minority class examples to properly learn the decision boundaries for the minority class) can best be addressed using a data-level method that acquires additional minority class training examples. But as we shall see in this chapter, sometimes such a direct solution is not available, and less direct methods must be utilized. Common misconceptions are also discussed and explained. Overall, this chapter provides an understanding of the foundations of imbalanced learning by providing a clear description of the relevant issues, and a clear mapping of these *issues* to the *methods* that can be used to address them.

Imbalanced Learning: Foundations, Algorithms, and Applications, First Edition.
Edited by Haibo He and Yunqian Ma.
© 2013 The Institute of Electrical and Electronics Engineers, Inc. Published 2013 by John Wiley & Sons, Inc.

2.1 INTRODUCTION

Many of the machine-learning and data-mining problems that we study, whether they are in business, science, medicine, or engineering, involve some form of data imbalance. The imbalance is often an integral part of the problem and in virtually every case the less frequently occurring entity is the one that we are most interested in. For example, those working on fraud detection will focus on identifying the fraudulent transactions rather than on the more common legitimate transactions [1], a telecommunications engineer will be far more interested in identifying the equipment about to fail than the equipment that will remain operational [2], and an industrial engineer will be more likely to focus on weld flaws than on welds that are completed satisfactorily [3].

In all these situations, it is far more important to accurately predict or identify the rarer case than the more common case, and this is reflected in the costs associated with errors in the predictions and classifications. For example, if we predict that telecommunication equipment is going to fail and it does not, we may incur some modest inconvenience and cost if the equipment is swapped out unnecessarily, but if we predict that equipment is not going to fail and it does, then we incur a much more significant cost when service is disrupted. In the case of medical diagnosis, the costs are even clearer: while a false-positive diagnosis may lead to a more expensive follow-up test and patient anxiety, a false-negative diagnosis could result in death if a treatable condition is not identified.

The authors of this chapter cover the foundations of imbalanced learning. It begins by providing important background information and terminology and then describes the fundamental issues associated with learning from imbalanced data. This description provides the foundation for understanding the imbalanced learning problem. This chapter then categorizes the methods for handling class imbalance and maps each to the fundamental issue that each method addresses. This mapping is quite important as many research papers on imbalanced learning fail to provide a comprehensive description of how or why these methods work, and what underlying issue(s) they address. This chapter provides a good overview of the imbalanced learning problem and describes some of the key work in the area, but it is not intended to provide either a detailed description of the methods used for dealing with imbalanced data or a comprehensive literature survey. Details on many of the methods are provided in subsequent chapters in this book.

2.2 BACKGROUND

A full appreciation of the issues associated with imbalanced data requires some important background knowledge. In this section, we look at what it means for a dataset to be imbalanced, what impact class imbalance has on learning, the role of between-class and within-class imbalances, and how imbalance applies to unsupervised learning tasks.

2.2.1 What is an Imbalanced Dataset and What is Its Impact on Learning?

We begin with a discussion of the most fundamental question: "What is meant by imbalanced data and imbalanced learning?" Initially, we focus on classification problems, and in this context, learning from imbalanced data means learning from data in which the classes have unequal numbers of examples. But because virtually no datasets are perfectly balanced, this is not a very useful definition. There is no agreement, or standard, concerning the exact degree of class imbalance required for a dataset to be considered truly "imbalanced." But most practitioners would certainly agree that a dataset where the most common class is less than twice as common as the rarest class would only be marginally unbalanced, that datasets with the imbalance ratio about 10:1 would be modestly imbalanced, and datasets with imbalance ratios above 1000:1 would be extremely unbalanced. But ultimately what we care about is how the imbalance impacts learning, and, in particular, the ability to learn the rare classes.

Learning performance provides us with an empirical—and objective—means for determining what should be considered an imbalanced dataset. Figure 2.1, generated from data in an earlier study that analyzed 26 binary-class datasets [4], shows how class imbalance impacts minority class classification performance. Specifically, it shows that the ratio between the minority class and the majority class error rates is greatest for the most highly imbalanced datasets and decreases as the amount of class imbalance decreases. Figure 2.1 clearly demonstrates that class imbalance leads to poorer performance when classifying minority class examples, as the error rate ratios are above 1.0. This impact is actually quite severe, as datasets with class imbalances between 5:1 and 10:1 have a minority class error rate more than 10 times that of the error rate on the majority class. The impact even appears quite significant for class imbalances between 1:1 and 3:1, which indicates that class imbalance is problematic in more situations than commonly acknowledged. This suggests that we should consider datasets with even moderate levels of class imbalance (e.g., 2:1) as "suffering" from class imbalance.

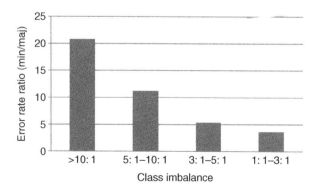

Figure 2.1 Impact of class imbalance on minority class performance.

There are a few subtle points concerning class imbalance. First, class imbalance must be defined with respect to a particular dataset or distribution. Since class labels are required in order to determine the degree of class imbalance, class imbalance is typically gauged with respect to the training distribution. If the training distribution is representative of the underlying distribution, as it is often assumed, then there is no problem; but if this is not the case, then we cannot conclude that the underlying distribution is imbalanced. But the situation can be complicated by the fact that when dealing with class imbalance, a common strategy is to artificially balance the training set. In this case, do we have class imbalance or not? The answer in this case is "yes"—we still do have class imbalance. That is, when discussing the problems associated with class imbalance, we really care about the underlying distribution. Artificially balancing the training distribution may help with the effects of class imbalance, but does not remove the underlying problem.

A second point concerns the fact that while class imbalance literally refers to the relative proportions of examples belonging to each class, the absolute number of examples available for learning is clearly very important. Thus, the class imbalance problem for a dataset with 10,000 positive examples and 1,000,000 negative examples is clearly quite different from a dataset with 10 positive examples and 1000 negative examples—even though the class proportions are identical. These two problems can be referred to as problems with relative rarity and absolute rarity. A dataset may suffer from neither of these problems, one of these problems, or both of these problems. We discuss the issue of absolute rarity in the context of class imbalance because highly imbalanced datasets very often have problems with absolute rarity.

2.2.2 Between-Class Imbalance, Rare Cases, and Small Disjuncts

Thus far we have been discussing class imbalance, or, as it has been termed, *between-class* imbalance. A second type of imbalance, which is not quite as well known or extensively studied, is *within-class* imbalance [5, 6]. Within-class imbalance is the result of rare cases [7] in the true, but generally unknown, classification concept to be learned. More specifically, rare cases correspond to sub-concepts in the induced classifier that covers relatively few cases. For example, in a medical dataset containing patient data where each patient is labeled as "sick" or "healthy," a rare case might correspond to those sick patients suffering from botulism, a relatively rare illness. In this domain, within-class imbalance occurs within the "sick" class because of the presence of much more general cases, such as those corresponding to the common cold. Just as the minority class in an imbalanced dataset is very hard to learn well, the rare cases are also hard to learn—even if they are part of the majority class. This difficulty is much harder to measure than the difficulty with learning the rare class, as rare cases can only be defined with respect to the classification concept, which, for real-world problems, is unknown, and can only be approximated. However, the difficulty of learning rare cases can be measured using artificial datasets that are generated

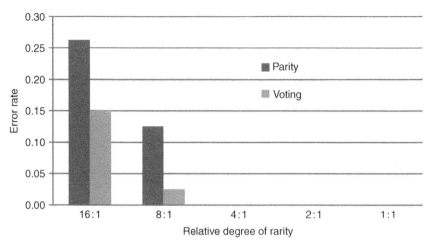

Figure 2.2 Impact of within-class imbalance on rare cases.

directly from a predefined concept. Figure 2.2 shows the results generated from the raw data from an early study on rare cases [7].

Figure 2.2 shows the error rate for the cases, or subconcepts, within the parity and voting datasets, based on how rare the case is relative to the most general case in the classification concept associated with the dataset. For example, a relative degree of rarity of 16 : 1 means that the rare case is 16 times as rare as the most common case, while a value of 1 : 1 corresponds to the most common case. For the two datasets shown in Figure 2.2, we clearly see that the rare cases (i.e., those with a higher relative degree of rarity) have a much higher error rate than the common cases, where, for this particular set of experiments, the more common cases are learned perfectly and have no errors. The concepts associated with the two datasets can be learned perfectly (i.e., there is no noise) and the errors were introduced by limiting the size of the training set.

Rare cases are difficult to analyze because one does not know the true concept and hence cannot identify the rare cases. This inability to identify these rare cases impacts the ability to develop strategies for dealing with them. But rare cases will manifest themselves in the learned concept, which is an approximation of the true concept. Many classifiers, such as decision tree and rule-based learners, form disjunctive concepts, and for these learners, the rare cases will form small disjuncts—the disjuncts in the learned classifier that cover few training examples [8]. The relationship between the rare and the common cases in the true (but generally unknown) concept, and the disjuncts in the induced classifier, is depicted in Figure 2.3.

Figure 2.3 shows a concept made up of two positively labeled cases, one is a rare case and the other is a common case, and the small and large disjuncts that the classifier forms to cover them. Any examples located within the solid boundaries corresponding to these two cases should be labeled as positive and

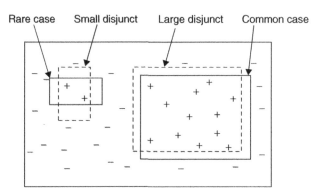

Figure 2.3   Relationship between rare/common cases and small/large disjuncts.

the data points outside these boundaries should be labeled as negative. The training examples are shown using the plus ("+") and the minus ("−") symbols. Note that the classifier will have misclassification errors on future test examples, as the boundaries for the rare and the common cases do not match the decision boundaries, represented by the dashed rectangles, which are formed by the classifier. Because approximately 50% of the decision boundary for the small disjunct falls outside the rare case, we expect this small disjunct to have an error rate near 50%. Applying similar reasoning, the error rate of the large disjunct in this case will only be about 10%. Because the uncertainty in this noise-free case mainly manifests itself near the decision boundaries, in such cases, we generally expect the small disjuncts to have a higher error rate, as a higher proportion of its "area" is near the decision boundary of the case to be learned. The difference between the induced decision boundaries and the actual decision boundaries in this case is mainly due to an insufficient number of training examples, although the bias of the learner also plays a role. In real-world situations, other factors, such as noise, will also have an effect.

The pattern of small disjuncts having a much higher error rates than large disjuncts, suggested by Figure 2.3, has been observed in practice in numerous studies [7–13]. This pattern is shown in Figure 2.4 for the classifier induced by C4.5 from the move dataset [13]. Pruning was disabled in this case as pruning has been shown to obscure the effect of small disjuncts on learning [12]. The disjunct size, specified on the x-axis, is determined by the number of training examples correctly classified by the disjunct (i.e., leaf node). The impact of the error-prone small disjuncts on learning is actually much greater than suggested by Figure 2.4, as the disjuncts of size 0–3, which correspond to the left-most bar in the figure, cover about 50% of the total examples and 70% of the errors.

In summary, we see that both rare classes and rare cases are difficult to learn and both lead to difficulties when learning from imbalanced data. When we discuss the foundational issues associated with learning from imbalanced data, we will see that these two difficulties are connected, in that, rare classes are disproportionately made up of rare cases.

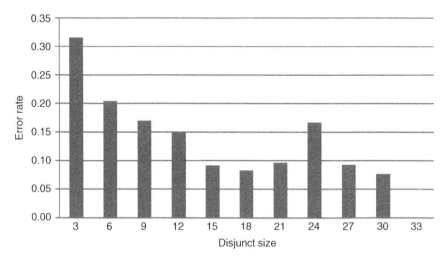

Figure 2.4 Impact of disjunct size on classifier performance (move dataset).

2.2.3 Imbalanced Data for Unsupervised Learning Tasks

Virtually all work that focuses explicitly on imbalanced data focuses on imbalanced data for classification. While classification is a key supervised learning task, imbalanced data can affect unsupervised learning tasks as well, such as clustering and association rule mining. There has been very little work on the effect of imbalanced data with respect to clustering, largely because it is difficult to quantify "imbalance" in such cases (in many ways, this parallels the issues with identifying rare cases). But certainly if there are meaningful clusters containing relatively few examples, existing clustering methods will have trouble identifying them. There has been more work in the area of association rule mining, especially with regard to market basket analysis, which looks at how the items purchased by a customer are related. Some groupings of items, such as *peanut butter* and *jelly*, occur frequently and can be considered common cases. Other associations may be extremely rare, but represent highly profitable sales. For example, *cooking pan* and *spatula* will be an extremely rare association in a supermarket, not because the items are unlikely to be purchased together, but because neither item is frequently purchased in a supermarket [14]. Association rule mining algorithms should ideally be able to identify such associations.

2.3 FOUNDATIONAL ISSUES

Now that we have established the necessary background and terminology, and demonstrated some of the problems associated with class imbalance, we are ready to identify and discuss the specific issues and problems associated with learning from imbalanced data. These issues can be divided into three major categories/levels: problem definition issues, data issues, and algorithm issues. Each of these categories is briefly introduced and then described in detail in subsequent sections.

Problem definition issues occur when one has insufficient information to properly define the learning problem. This includes the situation when there is no objective way to evaluate the learned knowledge, in which case one cannot learn an optimal classifier. Unfortunately, issues at the problem definition level are commonplace. Data issues concern the actual data that is available for learning and includes the problem of absolute rarity, where there are insufficient examples associated with one or more classes to effectively learn the class. Finally, algorithm issues occur when there are inadequacies in a learning algorithm that make it perform poorly for imbalanced data. A simple example involves applying an algorithm designed to optimize accuracy to an imbalanced learning problem where it is more important to classify minority class examples correctly than to classify majority class examples correctly.

2.3.1 Problem-Definition-Level Issues

A key task in any problem-solving activity is to *understand* the problem. As just one example, it is critically important for computer programmers to understand their customer's requirements before designing, and then implementing a software solution. Similarly, in data mining, it is critical for the data-mining practitioner to understand the problem and the user requirements. For classification tasks, this includes understanding how the performance of the generated classifier will be judged. Without such an understanding, it will be impossible to design an optimal or near-optimal classifier. While this need for evaluation information applies to all data-mining problems, it is particularly important for problems with class imbalance. In these cases, as noted earlier, the costs of errors are often asymmetric and quite skewed, which violates the default assumption of most classifier induction algorithms, which is that errors have uniform cost and thus accuracy should be optimized. The impact of using accuracy as an evaluation metric in the presence of class imbalance is well known—in most cases, poor minority class performance is traded off for improved majority class performance. This makes sense from an optimization standpoint, as overall accuracy is the weighted average of the accuracies associated with each class, where the weights are based on the proportion of training examples belonging to each class. This effect was clearly evident in Figure 2.1, which showed that the minority class examples have a much lower accuracy than majority class examples. What was not shown in Figure 2.1, but is shown by the underlying data [4], is that minority class predictions occur much less frequently than majority class predictions, even after factoring in the degree of class imbalance.

Accurate classifier evaluation information, if it exists, should be passed to the classifier induction algorithm. This can be done in many forms, one of the simplest forms being a cost matrix. If this information is available, then it is the algorithm's responsibility to utilize this information appropriately; if the algorithm cannot do this, then there is an algorithm-level issue. Fortunately, over the past decade, most classification algorithms have increased in sophistication so that they can handle evaluation criteria beyond accuracy, such as class-based misclassification costs and even costs that vary per example.

The problem definition issue also extends to unsupervised learning problems. Association rule mining systems do not have very good ways to evaluate the value of an association rule. But unlike the case of classification, as no single quantitative measure of quality is generated, this issue is probably better understood and acknowledged. Association rules are usually tagged with support and confidence values, but many rules with either high support or confidence values—or even both—will be uninteresting and potentially of little value. The lift of an association rule is a somewhat more useful measurement, but still does not consider the context in which the association will be used (lift measures how much more likely the antecedent and consequent of the rule are to occur together than if they were statistically independent). But as with classification tasks, imbalanced data causes further problems for the metrics most commonly used for association rule mining. As mentioned earlier, association rules that involve rare items are not likely to be generated, even if the rare items, when they do occur, often occur together (e.g., *cooking pan* and *spatula* in supermarket sales). This is a problem because such associations between rare items are more likely to be profitable because higher profit margins are generally associated with rare items.

2.3.2 Data-Level Issues

The most fundamental data-level issue is the lack of training data that often accompanies imbalanced data, which was previously referred to as an issue of *absolute rarity*. Absolute rarity does not only occur when there is imbalanced data, but is very often present when there are extreme degrees of imbalance—such as a class ratio of one to one million. In these cases, the number of examples associated with the rare class, or rare case, is small in an absolute sense. There is no predetermined threshold for determining absolute rarity and any such threshold would have to be domain specific and would be determined based on factors such as the dimensionality of the instance space, the distribution of the feature values within this instance space, and, for classification tasks, the complexity of the concept to be learned.

Figure 2.5 visually demonstrates the problems that can result from an "absolute" lack of data. The figure shows a simple concept, identified by the solid rectangle; examples within this rectangle belong to the positive class and examples outside this rectangle belong to the negative class. The decision boundary induced by the classifier from the labeled training data is indicated by the dashed rectangle. Figures 2.5a and 2.5b show the same concept but with Figure 2.5b having approximately half as many training examples as in Figure 2.5a. As one would expect, we see that the induced classifier more closely approximates the true decision boundary in Figure 2.5a because of the availability of additional training data.

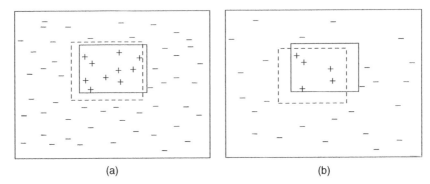

(a)   (b)

Figure 2.5 The impact of absolute rarity on classifier performance. (a) Full examples and (b) half of those used in (a).

Having a small amount of training data will generally have a much larger impact on the classification of the minority class (i.e., positive) examples. In particular, it appears that about 90% of the space associated with the positive class (in the solid rectangle) is covered by the learned classifier in Figure 2.5a, while only about 70% of it is covered in Figure 2.5b. One paper summarized this effect as follows: "A second reason why minority class examples are misclassified more often than majority class examples is that fewer minority class examples are likely to be sampled from the distribution D. Therefore, the training data are less likely to include (enough) instances of all of the minority class subconcepts in the concept space, and the learner may not have the opportunity to represent all truly positive regions. Because of this, some minority class test examples will be mistakenly classified as belonging to the majority class." [4, p. 325].

Absolute rarity also applies to rare cases, which may not contain sufficiently many training examples to be learned accurately. One study that used very simple artificially generated datasets found that once the training set dropped below a certain size, the error rate for the rare cases rose while the error rate for the general cases remained at zero. This occurred because with the reduced amount of training data, the common cases were still sampled sufficiently to be learned, but some of the rare cases were missed entirely [7]. The same study showed, more generally, that rare cases have a much higher misclassification rate than common cases. We refer to this as the *problem with rare cases*. This research also demonstrated something that had previously been assumed—that rare cases cause small disjuncts in the learned classifier. The *problem with small disjuncts*, observed in many empirical studies, is that they (i.e., small disjuncts) generally have a much higher error rate than large disjuncts [7–12]. This phenomenon is again the result of a lack of data. The most thorough empirical study of small disjuncts analyzed 30 real-world datasets and showed that, for the classifiers induced from these datasets, the vast majority of errors are concentrated in the smaller disjuncts [12].

These results suggest that absolute rarity poses a very serious problem for learning. But the problem could also be that small disjuncts sometimes do not represent rare, or exceptional, cases, but instead represent noise. The underlying problem, then, is that there is no easy way to distinguish between those small disjuncts that represent rare/exceptional cases, which should be kept, and those that represent noise, which should be discarded (i.e., pruned).

We have seen that rare cases are difficult to learn because of a lack of training examples. It is generally assumed that rare classes are difficult to learn for similar reasons. But in theory, it could be that rare classes are not disproportionately made up of rare cases, when compared to the makeup of common classes. But one study showed that this is most likely not the case as, across 26 datasets, the disjuncts labeled with the minority class were much smaller than the disjuncts with majority class labels [4]. Thus, rare classes tend to be made up of more rare cases (on the assumption that rare cases form small disjuncts) and as these are harder to learn than common cases, the minority class will tend to be harder to learn than the majority class. This effect is therefore due to an absolute lack of training examples for the minority class.

Another factor that may exacerbate any issues that already exist with imbalanced data is *noise*. While noisy data is a general issue for learning, its impact is magnified when there is imbalanced data. In fact, we expect noise to have a greater impact on rare cases than on common cases. To see this, consider Figure 2.6. Figure 2.6a includes no noisy data, while Figure 2.6b includes a few noisy examples. In this case, a decision tree classifier is used, which is configured to require at least two examples at the terminal nodes as a means of overfitting avoidance. We see that in Figure 2.6b, when one of the two training examples in the rare positive case is erroneously labeled as belonging to the negative class, the classifier misses the rare case completely, as two positive training examples are required to generate a leaf node. The less rare positive case, however, is not significantly affected because most of the examples in the induced disjunct are still positive and the two erroneously labeled training examples are not sufficient to alter the decision boundaries. Thus, noise will have a more significant impact on the rare cases than on the common cases. Another way to look at things is that it will be hard to distinguish between rare cases and noisy data points. Pruning, which is often used to combat noise, will remove the rare cases and the noisy cases together.

It is worth noting that while this section highlights the problem with absolute rarity, it does not highlight the problem with relative rarity. This is because we view relative rarity as an issue associated with the algorithm level. The reason is that class imbalance, which generally focuses on the relative differences in class proportions, is not fundamentally a problem at the data level—it is simply a property of the data distribution. We maintain that the problems associated with class imbalance and relative rarity are due to the lack of a proper problem formulation (with accurate evaluation criteria) or with algorithmic limitations with existing learning methods. The key point is that relative rarity/class imbalance is

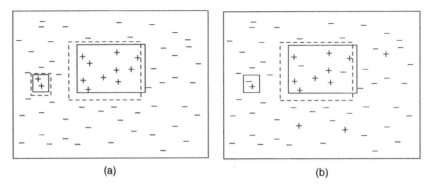

Figure 2.6 The effect of noise on rare cases. (a) No noisy data and (b) few noisy examples.

a problem only because learning algorithms cannot effectively handle such data. This is a very fundamental point, but one that is not often acknowledged.

2.3.3 Algorithm-Level Issues

There are a variety of algorithm-level issues that impact the ability to learn from imbalanced data. One such issue is the inability of some algorithms to optimize learning for the target evaluation criteria. Although this is a general issue with learning, it affects imbalanced data to a much greater extent than balanced data because in the imbalanced case, the evaluation criteria typically diverge much further from the standard evaluation metric—accuracy. In fact, most algorithms are still designed and tested much more thoroughly for accuracy optimization than for the optimization of other evaluation metrics. This issue is impacted by the metrics used to guide the heuristic search process. For example, decision trees are generally formed in a top–down manner and the tree construction process focuses on selecting the best test condition to expand the extremities of the tree. The quality of the test condition (i.e., the condition used to split the data at the node) is usually determined by the "purity" of a split, which is often computed as the weighted average of the purity values of each branch, where the weights are determined by the fraction of examples that follow that branch. These metrics, such as information gain, prefer test conditions that result in a balanced tree, where purity is increased for most of the examples, in contrast to test conditions that yield high purity for a relatively small subset of the data but low purity for the rest [15]. The problem with this is that a single high purity branch that covers only a few examples may identify a rare case. Thus, such search heuristics are biased against identifying highly accurate rare cases, which will also impact their performance on rare classes (which, as discussed earlier, often comprise rare cases).

The bias of a learning algorithm, which is required if the algorithm is to generalize from the data, can also cause problems when learning from imbalanced data. Most learners utilize a bias that encourages generalization and simple models to avoid the possibility of overfitting the data. But studies have shown that such biases work well for large disjuncts but not for small disjuncts [8], leading to the observed problem with small disjuncts (these biases tend to make the small disjuncts overly general). Inductive bias also plays a role with respect to rare classes. Many learners prefer the more common classes in the presence of uncertainty (i.e., they will be biased in favor of the class priors). As a simple example, imagine a decision tree learner that branches on all possible feature values when splitting a node in the tree. If one of the resulting branches covers no training examples, then there is no evidence on which to base a classification. Most decision tree learners will predict the most frequently occurring class in this situation, biasing the results against rarer classes.

The algorithm-level issues discussed thus far concern the use of search heuristics and inductive biases that favor the common classes and cases over the rare classes and cases. But the algorithm-level issues do not just involve favoritism. It is fundamentally more difficult for an algorithm to identify rare patterns than to identify relatively common patterns. There may be quite a few instances of the rare pattern, but the sheer volume of examples belonging to the more common patterns will obscure the relatively rare patterns. This is perhaps best illustrated with a variation of a common idiom in English: finding relatively rare patterns is "like finding needles in a haystack." The problem in this case is not so much that there are few needles, but rather that there is so much more hay.

The problem with identifying relatively rare patterns is partly due to the fact that these patterns are not easily located using the greedy search heuristics that are in common use. Greedy search heuristics have a problem with relative rarity because the rare patterns may depend on the conjunction of many conditions, and therefore examining any single condition in isolation may not provide much information or guidance. While this may also be true of common objects, with rare objects the impact is greater because the common objects may obscure the true signal. As a specific example of this general problem, consider the association rule mining problem described earlier, where we want to be able to detect the association between *cooking pan* and *spatula*. The problem is that both items are rarely purchased in a supermarket, so that even if the two are often purchased together when either one is purchased, this association may not be found. To find this association, the minimum support threshold for the algorithm would need to be set quite low. However, if this is done, there will be a combinatorial explosion because frequently occurring items will be associated with one another in an enormous number of ways. This association rule mining problem has been called the *rare item problem* [14] and it is an analog of the problem of identifying rare cases in classification problems. The fact that these random co-occurrences will swamp the meaningful associations between rare items is one example of the problem with relative rarity.

Another algorithm-level problem is associated with the divide-and-conquer approach that is used by many classification algorithms, including decision tree algorithms. Such algorithms repeatedly partition the instance space (and the examples that belong to these spaces) into smaller and smaller pieces. This process leads to data fragmentation [16], which is a significant problem when trying to identify rare patterns in the data, because there is less data in each partition from which to identify the rare patterns. Repeated partitioning can lead to the problem of absolute rarity within an individual partition, even if the original dataset exhibits only the problem of relative rarity. Data-mining algorithms that do not employ a divide-and-conquer approach therefore tend to be more appropriate when mining rare classes/cases.

2.4 METHODS FOR ADDRESSING IMBALANCED DATA

This section describes the methods that address the issues with learning from imbalanced data that were identified in the previous section. These methods are organized based on whether they operate at the problem definition, data, or algorithm level. As methods are introduced, the underlying issues that they address are highlighted. While this section covers most of the major methods that have been developed to handle imbalanced data, the list of methods is not exhaustive.

2.4.1 Problem-Definition-Level Methods

There are a number of methods for dealing with imbalanced data that operate at the problem definition level. Some of these methods are relatively straightforward, in that they directly address foundational issues that operate at the same level. But because of the inherent difficulty of learning from imbalanced data, some methods have been introduced that simplify the problem in order to produce more reasonable results. Finally, it is important to note that in many cases, there simply is insufficient information to properly define the problem and in these cases, the best option is to utilize a method that moderates the impact of this lack of knowledge.

2.4.1.1 Use Appropriate Evaluation Metrics It is always preferable to use evaluation metrics that properly factor in how the mined knowledge will be used. Such metrics are essential when learning from imbalanced data because they will properly value the minority class. These metrics can be contrasted with accuracy, which places more weight on the common classes and assigns value to each class proportional to its frequency in the training set. The proper solution is to use meaningful and appropriate evaluation metrics and for imbalanced data, this typically translates into providing accurate cost information to the learning algorithms (which should then utilize cost-sensitive learning to produce an appropriate classifier).

Unfortunately, it is not always possible to acquire the base information necessary to design good evaluation metrics that properly value the minority class. The next best solution is to provide evaluation metrics that are robust, given this lack of knowledge, where "robust" means that the metrics yield good results over a wide variety of assumptions. If these metrics are to be useful for learning from imbalanced datasets, they will tend to value the minority class much more than accuracy, which is now widely recognized as a poor metric when learning from imbalanced data. This recognition has led to the ascension of new metrics to replace accuracy for learning from unbalanced data.

A variety of metrics are routinely used when learning from imbalanced data when accurate evaluation information is not available. The most common metric involves receiver operation characteristics (ROC) analysis, and the area under the ROC curve (AUC) [17, 18]. ROC analysis can sometimes identify optimal models and discard suboptimal ones independent of the cost context or the class distribution (i.e., if one ROC curve dominates another), although in practice ROC curves tend to intersect, so that there is no one dominant model. ROC analysis does not have any bias toward models that perform well on the majority class at the expense of the majority class—a property that is quite attractive when dealing with imbalanced data. AUC summarizes this information into a single number, which facilitates model comparison when there is no dominating ROC curve. Recently, there has been some criticism concerning the use of ROC analysis for model comparison [19], but nonetheless this measure is still the most common metric used for learning from imbalanced data.

Other common metrics used for imbalanced learning are based on precision and recall. The precision of classification rules is essentially the accuracy associated with those rules, while the recall of a set of rules (or a classifier) is the percentage of examples of a designated class that are correctly predicted. For imbalanced learning, recall is typically used to measure the coverage of the minority class. Thus, precision and recall make it possible to assess the performance of a classifier on the minority class. Typically, one generates precision and recall curves by considering alternative classifiers. Just like AUC is used for model comparison for ROC analysis, there are metrics that combine precision and recall into a single number to facilitate comparisons between models. These include the geometric mean (the square root of precision times recall) and the F-measure [20]. The F-measure is parameterized and can be adjusted to specify the relative importance of precision versus recall, but the F1-measure, which weights precision and recall equally, is the variant most often used when learning from imbalanced data.

It is also important to use appropriate evaluation metrics for unsupervised learning tasks that must handle imbalanced data. As described earlier, association rule mining treats all items equally even though rare items are often more important than common ones. Various evaluation metrics have been proposed to deal with this imbalance and algorithms have been developed to mine association rules that satisfy these metrics. One simple metric assigns uniform weights to each item to represent its importance, perhaps its per-unit profit [21]. A slightly more sophisticated metric allows this weight to vary based on the transaction it appears in, which can be used to reflect the quantity of the item [22, 23]. But such measures still cannot represent simple metrics such as total profit. Utility mining [24, 25] provides this capability by allowing one to specify a uniform weight to represent per-item profit and a transaction weight to represent a quantity value. Objective-oriented association rule mining [26] methods, which make it possible to measure how well an association rule meets a user's objective, can be used to find association rules in a medical dataset where only treatments that have minimal side effects and minimum levels of effectiveness are considered.

2.4.1.2 Redefine the Problem   One way to deal with a difficult problem is to convert it into a simpler problem. The fact that the problem is not an equivalent problem may be outweighed by the improvement in results. This topic has received very little attention in the research community, most likely because it is not viewed as a research-oriented solution and is highly domain specific. Nonetheless, this is a valid approach that should be considered. One relatively general method for redefining a learning problem with imbalanced data is to focus on a subdomain or partition of the data, where the degree of imbalance is lessened. As long as this subdomain or partition is easily identified, this is a viable strategy. It may also be a more reasonable strategy than removing the imbalance artificially via sampling. As a simple example, in medical diagnosis, one could restrict the population to people over 90 years of age, especially if the targeted disease tends to be more common in the aged. Even if the disease occurs much more rarely in the young, using the entire population for the study could complicate matters if the people under 90, because of their much larger numbers, collectively contribute more examples of the disease. Thus, the strategy is to find a subdomain where the data is less imbalanced, but where the subdomain is still of sufficient interest. Other alternative strategies might be to group similar rare classes together and then simplify the problem by predicting only this "super-class."

2.4.2   Data-Level Methods

The main data-level issue identified earlier involves absolute rarity and a lack of sufficient examples belonging to rare classes and, in some cases, to the rare cases that may reside in either a rare or a common class. This is a very difficult issue to address, but methods for doing this are described in this section. This section also describes methods for dealing with relative rarity (the standard class imbalance problem), even though, as we shall discuss, we believe that issues with relative rarity are best addressed at the algorithms level.

2.4.2.1 Active Learning and Other Information Acquisition Strategies   The most direct way of addressing the issue of absolute rarity is to acquire additional labeled training data. Randomly acquiring additional labeled training data will be helpful and there are heuristic methods to determine whether the projected improvement in classification performance warrants the cost of obtaining more training data—and how many additional training examples should be acquired [27]. But a more efficient strategy is to preferentially acquire data from the rare classes or rare cases. Unfortunately, this cannot easily be done directly as one cannot identify examples belonging to rare classes and rare cases with certainty. But there is an expectation that active learning strategies will tend to preferentially sample such examples. For example, uncertainty sampling methods [28] are likely to focus more attention on rare cases, which will generally yield less certain predictions because of the smaller number of training examples to generalize from. Put another way, as small disjuncts have a much higher error rate than large disjuncts, it seems clear that active learning methods would focus on obtaining examples belonging to those disjuncts. Other work on active learning has further demonstrated that active learning methods are capable of preferentially sampling the rare classes by focusing the learning on the instances around the classification boundary [29]. This general information acquisition strategy is supported by the empirical evidence that shows that balanced class distributions generally yield better performance than unbalanced ones [4].

Active learning and other simpler information acquisition strategies can also assist with the relative rarity problem, as such strategies, which acquire examples belonging to the rarer classes and rarer cases, address the relative rarity problem while addressing the absolute rarity problem. Note that this is true even if uncertainty sampling methods tend to acquire examples belonging to rare cases, as prior work has shown that rare cases tend to be more associated with the rarer classes [4]. In fact, this method for dealing with relative rarity is to be preferred to the sampling methods addressed next, as those methods do not obtain new knowledge (i.e., valid new training examples).

2.4.2.2 Sampling Methods Sampling methods are a very popular method for dealing with imbalanced data. These methods are primarily employed to address the problem with relative rarity but do not address the issue of absolute rarity. This is because, with the exception of some methods that utilize some intelligence to generate new examples, these methods do not attack the underlying issue with absolute rarity—a lack of examples belonging to the rare classes and rare cases. But, as will be discussed in Section 2.4.3, our view is also that sampling methods do not address the underlying problem with relative rarity either. Rather, sampling masks the underlying problem by artificially balancing the data, without solving the basic underlying issue. The proper solution is at the algorithm level and requires algorithms that are designed to handle imbalanced data.

The most basic sampling methods are random undersampling and random oversampling. Random undersampling randomly eliminates majority class examples from the training data, while random oversampling randomly duplicates minority class training examples. Both of these sampling techniques decrease the degree of class imbalance. But as no new information is introduced, any underlying issues with absolute rarity are not addressed. Some studies have shown random oversampling to be ineffective at improving recognition of the minority class [30, 31], while another study has shown that random undersampling is ineffective [32]. These two sampling methods also have significant drawbacks. Undersampling discards potentially useful majority class examples, while oversampling increases the time required to train a classifier and also leads to overfitting that occurs to cover the duplicated training examples [31, 33].

More advanced sampling methods use some intelligence when removing or adding examples. This can minimize the drawbacks that were just described and, in the case of intelligently adding examples, has the potential to address the underlying issue of absolute rarity. One undersampling strategy removes only majority class examples that are redundant with other examples or border regions with minority class examples, figuring that they may be the result of noise [34]. Synthetic minority oversampling technique (SMOTE), on the other hand, oversamples the data by introducing new, non-replicated minority class examples from the line segments that join the five minority class nearest neighbors [33]. This tends to expand the decision boundaries associated with the small disjuncts/rare cases, as opposed to the overfitting associated with random oversampling. Another approach is to identify a good class distribution for learning and then generate samples with that distribution. Once this is done, multiple training sets with the desired class distribution can be formed using all minority class examples and a subset of the majority class examples. This can be done so that each majority class example is guaranteed to occur in at least one training set; so no data is wasted. The learning algorithm is then applied to each training set and meta-learning is used to form a composite learner from the resulting classifiers. This approach can be used with any learning method and it was applied to four different learning algorithms [1]. The same basic approach for partitioning the data and learning multiple classifiers has also been used with support vector machines and an support vector machine (SVM) ensemble has outperformed both undersampling and oversampling [35].

All of these more sophisticated methods attempt to reduce some of the drawbacks associated with the simple random sampling methods. But for the most part, it seems unlikely that they introduce any new knowledge and hence they do not appear to truly address any of the underlying issues previously identified. Rather, they at best compensate for learning algorithms that are not well suited to dealing with class imbalance. This point is made quite clearly in the description of the SMOTE method, when it is mentioned that the introduction of the new examples effectively serves to change the *bias* of the learner, forcing a more general bias, but only for the minority class. Theoretically, such a modification to the bias could be implemented at the algorithm level. As discussed later, there has been research at the algorithm level in modifying the bias of a learner to better handle imbalanced data.

The sampling methods just described are designed to reduce between-class imbalance. Although research indicates that reducing between-class imbalance will also tend to reduce within-class imbalances [4], it is worth considering whether sampling methods can be used in a more direct manner to reduce within-class imbalances—and whether this is beneficial. This question has been studied using artificial domains and the results indicate that it is not sufficient to eliminate between-class imbalances (i.e., rare classes) in order to learn complex concepts that contain within-class imbalances (i.e., rare cases) [5]. Only when the within-class imbalances are also eliminated, the concept can be learned well. This suggests that sampling should be used to improve the performance associated with rare cases. Unfortunately, there are problems with implementing the strategy for real-world domains where one cannot identify the rare cases. The closest we can get to this approach is to assume that rare cases correspond to small disjuncts in the induced classifier and then sample based on disjunct size, with the goal of equalizing the sizes of the disjuncts in the induced classifier.

2.4.3 Algorithm-Level Methods

A number of algorithm-level methods have been developed to handle imbalanced data. The majority of these techniques involve using search methods that are well suited for identifying rare patterns in data when common patterns abound.

2.4.3.1 Search Methods that Avoid Greed and Recursive Partitioning Greedy search methods and search methods that use a divide-and-conquer approach to recursively partition the search space have difficulty in finding rare patterns, for the reasons provided in Section 2.3.3. Thus, learning methods that avoid, or minimize, these two approaches will tend to perform better when there is imbalanced data. The advances in computational power that have occurred since many of the basic learning methods were introduced make it more practical to utilize less greedy search heuristics. Perhaps the best example of this is genetic algorithms, which are global search techniques that work with populations of candidate solutions rather than a single solution and employ stochastic operators to guide the search process [36]. These methods tend to be far less greedy than many popular learning methods and these characteristics permit genetic algorithms to cope well with attribute interactions [36, 37] and avoid getting stuck in local maxima, which together make genetic algorithms very suitable for dealing with rarity. In addition, genetic algorithms also do not rely on a divide-and-conquer approach that leads to the data fragmentation problem. Several systems have relied on the power of genetic algorithms to handle rarity. Weiss [38] uses a genetic algorithm to predict very rare events, while Carvalho and Freitas [39, 40] use a genetic algorithm to discover "small disjunct rules." Certainly other search methods are less greedy than decision trees and also do not suffer from the data fragmentation problems. However, no truly comprehensive study has examined a large variety of different search methods over a large variety of imbalanced datasets; so definitive conclusions cannot be drawn. Such studies would be useful and it is interesting to note that these types of large-scale empirical studies have been conducted to compare the effectiveness of sampling methods—which have garnered much more focused attention from the imbalanced data community.

2.4.3.2 Search Methods that Use Metrics Designed to Handle Imbalanced Data One problem-level method for handling class imbalance involves using evaluation metrics that properly value the learned/mined knowledge. However, evaluation metrics also play a role at the algorithm level to guide the heuristic search process. Some metrics have been developed to improve this search process when dealing with imbalanced data—most notably metrics based on precision and recall. Search methods that focus on simultaneously maximizing precision and recall may fail because of the difficulty of optimizing these competing values; so some systems adopt more sophisticated approaches. Timeweaver [38], a genetic algorithm-based classification system, periodically modifies the parameter to the F-measure that controls the relative importance of precision and recall in the fitness function, so that a diverse set of classification rules is evolved, with some rules having high precision and others high recall. The expectation is that this will eventually lead to rules with both high precision and recall. A second approach optimizes recall in the first phase of the search process and precision in the second phase by eliminating false positives covered by the rules [41]. Returning to the needle and haystack analogy, this approach identifies regions likely to contain needles in the first phase and then discards strands of hay within these regions in the second phase.

2.4.3.3 *Inductive Biases Better Suited for Imbalanced Data*   Most inductive learning systems heavily favor generality over specialization. While an inductive bias that favors generality is appropriate for learning common cases, it is not appropriate for rare cases and may even cause rare cases to be totally ignored. There have been several attempts to improve the performance of data-mining systems with respect to rarity by choosing a more appropriate bias. The simplest approach involves modifying existing systems to eliminate some small disjuncts based on tests of statistical significance or using error estimation techniques—often as a part of an overfitting avoidance strategy. The hope is that these will remove only improperly learned disjuncts, but such methods will also remove those disjuncts formed to cover rare cases. The basic problem is that the significance of small disjuncts cannot be reliably estimated and consequently significant small disjuncts may be eliminated along with the insignificant ones. Error estimation techniques are also unreliable when there are only a few examples, and hence they suffer from the same basic problem. These approaches work well for large disjuncts because in these cases statistical significance and error rate estimation techniques yield relatively reliable estimates—something they do not do for small disjuncts.

More sophisticated approaches have been developed but the impact of these strategies on rare cases cannot be measured directly, as the rare cases in the true concept are generally not known. Furthermore, in early work on this topic, the focus was on the performance of small disjuncts; so it is difficult to assess the impact of these strategies on class imbalance. In one study, the learner's maximum generality bias was replaced with a maximum specificity bias for the small disjuncts, which improved the performance of the small disjuncts but degraded the performance of the larger disjuncts and the overall accuracy [8]. Another study also utilized a maximum specificity bias but took steps to ensure that this did not impact the performance of the large disjuncts by using a different learning method classify them [11]. A similar hybrid approach was also used in one additional study [39, 40].

Others advocate the use of instance-based learning for domains with many rare cases/small disjuncts because of the highly specific bias associated with this learning method [10]. In such methods, all training examples are generally stored in memory and utilized, as compared to other approaches where examples, when they fall below some utility threshold, are ignored (e.g., because of pruning). In summary, there have been several attempts to select an inductive bias that will perform better in the presence of small disjuncts that are assumed to represent rare cases. But these methods have shown only mixed success and, most significantly, this work has not directly examined the class imbalance; these methods may assist with the class imbalance as rare classes are believed to be formed disproportionately from rare cases. Such approaches, which have not garnered much attention in the past decade, are quite relevant and should be reexamined in the more modern context of class imbalance.

2.4.3.4 Algorithms that Implicitly or Explicitly Favor Rare Classes and Cases

Some algorithms preferentially favor the rare classes or cases and hence tend to perform well on classifying rare classes and cases. Cost-sensitive learning algorithms are one of the most popular such algorithms for handling imbalanced data. While the assignment of costs in response to the problem characteristics is done at the problem level, cost-sensitive learning must ultimately be implemented at the algorithm level. There are several algorithmic methods for implementing cost-sensitive learning, including weighting the training examples in a cost-proportionate manner [42] and building the cost sensitivity directly into the learning algorithm [43]. These iterative algorithms place different weights on the training distribution after each iteration and increase (decrease) the weights associated with the incorrectly (correctly) classified examples. Because rare classes/cases are more error-prone than common classes/cases [4, 38], it is reasonable to believe that boosting will improve their classification performance. Note that because boosting effectively alters the distribution of the training data, one could consider it a type of advanced adaptive sampling technique. AdaBoost's weight-update rule has also been made cost sensitive, so that misclassified examples belonging to rare classes are assigned higher weights than those belonging to common classes. The resulting system, Adacost [44], has been empirically shown to produce lower cumulative misclassification costs than AdaBoost and thus, like other cost-sensitive learning methods, can be used to address the problem with rare classes.

Boosting algorithms have also been developed to directly address the problem with rare classes. RareBoost [45] scales false-positive examples in proportion to how well they are distinguished from true-positive examples and scales false-positive examples in proportion to how well they are distinguished from true-negative examples. A second algorithm that uses boosting to address the problems with rare classes is SMOTEBoost [46]. This algorithm recognizes that boosting may suffer from the same problems as oversampling (e.g., overfitting), as boosting will tend to weight examples belonging to the rare classes more than those belonging to the common classes—effectively duplicating some of the examples belonging to the rare classes. Instead of changing the distribution of training data by updating the weights associated with each example, SMOTEBoost alters the distribution by adding new minority class examples using the SMOTE algorithm [33].

2.4.3.5 Learn Only the Rare Class   The problem of relative rarity often causes the rare classes to be ignored by classifiers. One method of addressing this data-level problem is to employ an algorithm that only learns classification rules for the rare class, as this will prevent the more common classes from overwhelming the rarer classes. There are two main variations to this approach. The recognition-based approach learns only from examples associated with the rare class, thus recognizing the patterns shared by the training examples, rather than discriminating between examples belonging to different classes. Several systems have used such recognition-based methods to learn rare classes [47, 48].

The other approach, which is more common and supported by several learning algorithms, learns from examples belonging to all classes but first learns rules to cover the rare classes [15, 49, 50]. Note that this approach avoids most of the problems with data fragmentation, as examples belonging to the rare classes will not be allocated to the rules associated with the common classes before any rules are formed that cover the rare classes. Such methods are also free to focus only on the performance of the rules associated with the rare class and not worry about how this affects the overall performance of the classifier [15, 50]. Probably the most popular such algorithm is the Ripper algorithm [49], which builds rules using a separate-and-conquer approach. Ripper normally generates rules for each class from the rarest class to the most common class. At each stage, it grows rules for the one targeted class by adding conditions until no examples are covered, which belong to the other classes. This leads to highly specialized rules, which are good for covering rare cases. Ripper then covers the most common class using a default rule that is used when no other rule is applicable.

2.4.3.6 Algorithms for Mining Rare Items   Association rule mining is a well-understood area. However, when metrics other than support and confidence are used to identify item sets or their association rules, algorithmic changes are required. In Section 2.4.1, we briefly discussed a variety of metrics for finding association rules when additional metrics are added to support and confidence. We did not describe the corresponding changes to the association rule mining algorithms, but they are described in detail in the relevant papers [21–26].

There is also an algorithmic solution to the rare item problem, in which significant associations between rarely occurring items may be missed because the minimum support value *minsup* cannot be set too low, as a very low value would cause a combinatorial explosion of associations. This problem can be solved by specifying multiple minimum levels of support to reflect the frequencies of the associated items in the distribution [14]. Specifically, the user can specify a different *minsup* value for each item. The minimum support for an association rule is then the lowest *minsup* value among the items in the rule. Association rule mining systems are tractable mainly because of the downward closure property of support: if a set of items satisfies *minsup*, then so do all of its subsets. While this downward closure property does not hold with multiple minimum levels of support, the standard Apriori algorithm for association rule mining can be modified to satisfy the sorted closure property for multiple minimum levels of support [14]. The use of multiple minimum levels of support then becomes tractable. Empirical results indicate that the new algorithm is able to find meaningful associations involving rare items without producing a huge number of meaningless rules involving common items.

2.5 MAPPING FOUNDATIONAL ISSUES TO SOLUTIONS

This section summarizes the foundational problems with imbalanced data described in Section 2.3 and how they can be addressed by the various methods described in Section 2.4. This section is organized using the three basic categories identified earlier in this chapter: problem definition level, data level, and algorithm level.

The problem-definition-level issues arise because researchers and practitioners often do not have all of the necessary information about a problem to solve it optimally. Most frequently this involves not possessing the necessary metrics to accurately assess the utility of the mined knowledge. The solution to this problem is simple, although often not achievable: obtain the requisite knowledge and from this generate the metrics necessary to properly evaluate the mined knowledge. Because this is not often possible, one must take the next best course of action—use the best available metric or one that is at least "robust" such that it will lead to good, albeit suboptimal solutions, given incomplete knowledge and hence inexact assumptions. In dealing with imbalanced data, this often means using ROC analysis when the necessary evaluation information is missing. One alternate solution that was briefly discussed involves redefining the problem to a simpler problem for which more exact evaluation information is available. Fortunately the state of the art in data- mining technology has advanced to the point where in most cases if we do have the precise evaluation information, we can utilize it; in the past, data-mining algorithms were often not sufficiently sophisticated to incorporate such knowledge.

Data-level issues also arise when learning from imbalanced data. These issues mainly relate to absolute rarity. Absolute rarity occurs when one or more classes do not have sufficient numbers of examples to adequately learn the decision boundaries associated with that class. Absolute rarity has a much bigger impact on the rare classes than on common classes. Absolute rarity also applies to rare cases, which may occur for either rare classes or common classes, but are disproportionately associated with rare classes. The ideal and most straightforward approach to handling absolute rarity, in either of its two main forms, is to acquire additional training examples. This can often be done most efficiently via active learning and other information acquisition strategies.

It is important to understand that we do not view class imbalance, which results from a relative difference in frequency between the classes, as a problem at the data level—the problem only exists because most algorithms do not respond well to such imbalances. The straightforward method for dealing with class imbalance is via sampling, a method that operates at the data level. But this method for dealing with class imbalance has many problems, as we discussed previously (e.g., undersampling involves discarding potentially useful data) and is far from ideal. A much better solution would be to develop algorithms that can handle the class imbalance. At the current moment, sampling methods do perform competitively and therefore cannot be ignored, but it is important to recognize that such methods will always have limited value and that algorithmic solutions can potentially be more effective. We discuss these methods next (e.g., one-class learning) because we view them as addressing foundational algorithmic issues.

Algorithm-level issues mainly involve the ability to find subtle patterns in data that may be obscured because of imbalanced data and class imbalance, in particular (i.e., relative rarity). Finding patterns, such as those that identify examples belonging to a very rare class, is a very difficult task. To accomplish this task, it is important to have an appropriate search algorithm, a good evaluation metric to guide the heuristic search process, and an appropriate inductive bias. It is also important to deal with issues such as data fragmentation, which can be problematic especially for imbalanced data. The most common mechanism for dealing with this algorithm-level problem is to use sampling, a data-level method, to reduce the degree of class imbalance. But for reasons outlined earlier, this strategy does not address the foundational underlying issue—although it does provide some benefit. The strategies that function at the algorithm level include using a non-greedy search algorithm and one that does not repeatedly partition the search space; using search heuristics that are guided by metrics that are appropriate for imbalanced data; using inductive biases that are appropriate for imbalanced data; and using algorithms that explicitly or implicitly focus on the rare classes or rare cases, or only learn the rare class.

2.6 MISCONCEPTIONS ABOUT SAMPLING METHODS

Sampling methods are the most common methods for dealing with imbalanced data, but yet there are widespread misconceptions related to these methods. The most basic misconception concerns the notion that sampling methods are *equivalent* to certain other methods for dealing with class imbalance. In particular, Breiman et al. [51] establishes the connection between the distribution of training-set examples, the costs of mistakes on each class, and the placement of the decision threshold. Thus, for example, one can make false negatives twice as costly as false positives by assigning appropriate costs or by increasing the ratio of positive to negative examples in the training set by a factor of 2, or by setting the probability threshold for determining the class label to two-thirds rather than to one-half. Unfortunately, as implemented in real-world situations, these equivalences do *not* hold.

As a concrete example, suppose that a training set has 10,000 examples and a class distribution of 100 : 1, so that there are only 100 positive examples. One way to improve the identification of the rare class is to impose a greater cost for false negatives than for false positives. A cost ratio of 100 : 1 is theoretically equivalent to modifying the training distribution, so that it is balanced, with a 1 : 1 class ratio. To generate such a balanced distribution in practice, one would typically oversample the minority class or undersample the majority class, or do both. But if one undersamples the majority class, then potentially valuable data is thrown away, and if one oversamples the minority class, then one is making exact copies of examples, which can lead to overfitting. For the equivalence to hold, one should randomly select *new* minority class examples from the original distribution, which would include examples that are not already available for training. But this is almost never feasible. Even generating new, synthetic, minority class examples violates the equivalence, as these examples will, at best, only be a better *approximation* of the true distribution. Thus, sampling methods are not equivalent in practice to other methods for dealing with imbalanced data and they have drawbacks that other methods, such as cost-sensitive learning, do not have, if implemented properly.

Another significant concern with sampling is that its impact is often not fully understood—or even considered. Increasing the proportion of examples belonging to the rare class has two distinct effects. First, it will help address the problems with relative rarity, and, if the examples are new examples, will also address the problem with absolute rarity by injecting new knowledge. However, if no corrective action is taken, it will also have a second effect—it will impose nonuniform error costs, causing the learner to be biased in favor of predicting the rare class. In many situations, this second effect is desired and is the actually the main reason for altering the class distribution of the training data. But in other cases, namely when new examples are added (e.g., via active learning), this effect is not desirable. That is, in these other cases, the intent is to improve performance with respect to the rare class by having *more data* available for that class, not by biasing the data-mining algorithm toward that class. In these cases, this bias should be removed.

The bias introduced toward predicting the oversampled class can be removed using the equivalences noted earlier to account for the differences between the training distribution and the underlying distribution [4, 43]. For example, the bias can be removed by adjusting the decision thresholds, as was done in one study that demonstrated the positive impact of removing this unintended bias [4]. That study showed that adding new examples to alter the class distribution of the training data, so that it deviates from the natural, underlying, distribution, improved classifier performance. However, classifier performance was improved even more when the bias just described was removed by adjusting the decision thresholds within the classifier. Other research studies that investigate the use of sampling to handle rare cases and class imbalance almost never remove this bias—and worse yet, do not even discuss the implications of this decision. This issue must be considered much more carefully in future studies.

2.7 RECOMMENDATIONS AND GUIDELINES

The authors of this chapter categorized some of the major issues with imbalanced data and then described the methods most appropriate for handling each type of issue. Thus, one recommendation is to try to use those methods for handling imbalanced data that are most appropriate for dealing with the underlying issue. This usually means utilizing methods at the same level as the issue, when possible. But often the ideal method is not feasible—like using active learning to obtain more training data when there is an issue of absolute rarity. Thus, one must often resort to sampling, but in such cases, one should be aware of the drawbacks associated with these methods and avoid the common misconceptions associated with these methods. Unfortunately, it is not easy to effectively deal with imbalanced data because of the fundamental issues that are involved—which is probably why even after more than a decade of intense scrutiny, the research community still has much work remaining to come up with effective methods for dealing with these problems. Even methods that had become accepted, such as the use of AUC to generate robust classifiers when good evaluation metrics are not available, are now coming into question [19]. Nonetheless, there has been progress and certainly there is a much better appreciation of the problem than in the past.

REFERENCES

1. P. Chan and S. Stolfo, "Toward scalable learning with non-uniform class and cost distributions: A case study in credit card fraud detection," in *Proceedings of the Fourth International Conference on Knowledge Discovery and Data Mining* (New York, NY, USA), pp. 164–168, AAAI Press, 2001.
2. G. Weiss and H. Hirsh, "Learning to predict rare events in event sequences," in *Proceedings of the Fourth International Conference on Knowledge Discovery and Data Mining* (New York, NY, USA), pp. 359–363, AAAI Press, 1998.
3. T. Liao, "Classification of weld flaws with imbalanced data," *Expert Systems with Applications: An International Journal*, vol. 35, no. 3, pp. 1041–1052, 2008.
4. G. Weiss and F. Provost, "Learning when training data are costly: The effect of class distribution on tree induction," *Journal of Artificial Intelligence Research*, vol. 19, pp. 315–354, 2003.
5. N. Japkowicz, "Concept learning in the presence of between-class and within-class imbalances," in *Proceedings of the Fourteenth Conference of the Canadian Society for Computational Studies of Intelligence* (Ottawa, Canada), pp. 67–77, Springer-Verlag, 2001.

6. N. Japkowicz, "Class imbalances versus small disjuncts," *ACM SIGKDD Explorations*, vol. 6, no. 1, pp. 40–49, 2004.

7. G. Weiss, "Learning with rare cases and small disjuncts," in *Proceedings of the Twelfth International Conference on Machine Learning* (Tahoe City, CA, USA), pp. 558–565, Morgan Kaufmann, 1995.

8. R. Holte, L. Acker, and B. Porter, "Concept learning and the problem of small disjuncts," in *Proceedings of the Eleventh International Joint Conference on Artificial Intelligence* (Detroit, MI, USA), pp. 813–818, Morgan Kaufmann, 1989.

9. K. Ali and M. Pazzani, "HYDRA-MM: Learning multiple descriptions to improve classification accuracy," *International Journal of Artificial Intelligence Tools*, vol. 4, pp. 97–122, 1995.

10. A. van den Bosch, T. Weijters, H. J. van den Herik, and W. Daelemans, "When small disjuncts abound, try lazy learning: A case study," in *Proceedings of the Seventh Belgian-Dutch Conference on Machine Learning* (Tilburg, Netherlands), pp. 109–118, Tilburg University, 1997.

11. K. Ting, "The problem of small disjuncts: Its remedy in decision trees," in *Proceedings of the Tenth Canadian Conference on Artificial Intelligence*, pp. 91–97, Morgan Kaufmann, 1994.

12. G. Weiss and H. Hirsh, "A quantitative study of small disjuncts," in *Proceedings of the Seventeenth National Conference on Artificial Intelligence* (Austin, TX, USA), pp. 665–670, AAAI Press, 2000.

13. G. Weiss and H. Hirsh, "A quantitative study of small disjuncts: Experiments and results," Tech. Rep. ML-TR-42, Rutgers University, 2000.

14. B. Liu, W. Hsu, and Y. Ma, "Mining association rules with multiple minimum supports," in *Proceedings of the Fifth ACM SIGKDD International Conference on Knowledge Discovery and Data Mining* (San Diego, CA, USA), pp. 337–341, ACM, 1999.

15. P. Riddle, R. Segal, and O. Etzioni, "Representation design and brute-force induction in a Boeing manufacturing design," *Applied Artificial Intelligence*, vol. 8, pp. 125–147, 1994.

16. J. Friedman, R. Kohavi, and Y. Yun, "Lazy decision trees," in *Proceedings of the Thirteenth National Conference on Artificial Intelligence* (Portland, OR, USA), pp. 717–724, AAAI Press, 1996.

17. A. Bradley, "The use of the area under the ROC curve in the evaluation of machine learning algorithms," *Pattern Recognition*, vol. 30, no. 7, pp. 1145–1159, 1997.

18. F. Provost and T. Fawcett, "Robust classification for imprecise environments," *Machine Learning*, vol. 42, pp. 203–231, 2001.

19. D. Hand, "Measuring classifier performance: A coherent alternative to the area under the ROC curve," *Machine Learning*, vol. 77, pp. 103–123, 2009.

20. C. van Rijsbergen, *Information Retrieval*. London: Butterworths, 1979.

21. C. Cai, A. Fu, C. Cheng, and W. Kwong, "Mining association rules with weighted items," in *Proceedings of Database Engineering and Applications Symposium* (Cardiff, UK), pp. 68–77, IEEE Computer Society, 1998.

22. C. Carter, H. Hamilton, and J. Cercone, "Share based measures for itemsets," in *Principles of Data Mining and Knowledge Discovery*, Lecture Notes in Computer Science (J. Komorowski and J. Zytkow, eds.), vol. 1263, pp. 14–24, Berlin, Heidelberg/New York: Springer-Verlag, 1997.

23. W. Wang, J. Yang, and P. Yu, "Efficient mining of weighted association rules (WAR)," in *Proceedings of the 6th ACM International Conference on Knowledge Discovery and Data Mining* (Boston, MA, USA), pp. 270–274, ACM, 2000.

24. H. Yao, H. Hamilton, and C. Butz, "A foundational approach to mining itemset utilities from databases," in *Proceedings of the Fourth SIAM International Conference on Data Mining* (Lake Buena Vista, FL, USA), pp. 482–496, Society for Industrial and Applied Mathematics (SIAM), 2004.

25. J. Yao and J. Hamilton, "Mining itemset utilities from transaction databases," *Data and Knowledge Engineering*, vol. 59, no. 3, pp. 603–626, 2006.

26. Y. Shen, Q. Yang, and Z. Zhang, "Objective-oriented utility-based association mining," in *Proceedings of the 2002 IEEE International Conference on Data Mining* (Maebashi City, Japan), pp. 426–433, IEEE Computer Society, 2002.

27. G. Weiss and Y. Tian, "Maximizing classifier utility when there are data acquisition and modeling costs," *Data Mining and Knowledge Discovery*, vol. 17, no. 2, pp. 253–282, 2008.

28. D. Lewis and J. Catlett, "Heterogeneous uncertainty sampling for supervised learning," in *Proceedings of the Eleventh International Conference on Machine Learning* (New Brunswick, NJ, USA), pp. 148–156, Morgan Kaufmann, 1994.

29. S. Ertekin, J. Huang, and C. Giles, "Active learning for class imbalance problem," in *Proceedings of the 30th International Conference on Research and Development in Information Retrieval* (Amsterdam, The Netherlands), ACM, 2007.

30. C. Ling and C. Li, "Data mining for direct marketing problems and solutions," in *Proceedings of the Fourth International Conference on Knowledge Discovery and Data Mining* (New York, NY, USA), pp. 73–79, AAAI Press, 1998.

31. C. Drummond and R. Holte, "C4.5, class imbalance, and cost sensitivity: Why undersampling beats over-sampling," in *ICML Workshop on Learning from Imbalanced Data Sets II,*, (Washington, DC, USA), 2003.

32. N. Japkowicz and S. Stephen, "The class imbalance problem: A systematic study," *Intelligent Data Analysis*, vol. 6, no. 5, pp. 429–450, 2002.

33. N. Chawla, K. Bowyer, L. Hall, and W. Kegelmeyer, "SMOTE: Synthetic minority over-sampling technique," *Journal of Artificial Intelligence Research*, vol. 16, pp. 321–357, 2002.

34. M. Kubat and S. Matwin, "Addressing the curse of imbalanced training sets: One-sided selection," in *Proceedings of the Fourteenth International Conference on Machine Learning* (Nashville, TN, USA), pp. 179–186, Morgan Kaufmann, 1997.

35. R. Yan, Y. Liu, R. Jin, and A. Hauptmann, "On predicting rare classes with SVM ensembles in scene classification," in *Proceedings of IEEE International Conference on Acoustics, Speech and Signal Processing* (Hong Kong), IEEE Signal Processing Society, 2003.

36. D. Goldberg, *Genetic Algorithms in Search, Optimization and Machine Learning*. Reading, MA: Addison-Wesley, 1989.

37. A. Freitas, "Evolutionary computation," in *Handbook of Data Mining and Knowledge Discovery* (W. Klösgen and Jan M. Zytkow, eds.), pp. 698–706, New York, NY: Oxford University Press, 2002.

38. G. Weiss, "Timeweaver: A genetic algorithm for identifying predictive patterns in sequences of events," in *Proceedings of the Genetic and Evolutionary Computation Conference* (Orlando, FL, USA), pp. 718–725, Morgan Kaufmann, 1999.

39. D. Carvalho and A. A. Freitas, "A genetic algorithm for discovering small-disjunct rules in data mining," *Applied Soft Computing*, vol. 2, no. 2, pp. 75–88, 2002.
40. D. Carvalho and A. A. Freitas, "New results for a hybrid decision tree/genetic algorithm for data mining," in *Proceedings of the Fourth International Conference on Recent Advances in Soft Computing* (Nottingham, UK), pp. 260–265, Nottingham Trent University, 2002.
41. M. Joshi, R. Agarwal, and V. Kumar, "Mining needles in a haystack: Classifying rare classes via two-phase rule induction," in *SIGMOD '01 Conference on Management of Data* (Santa Barbara, CA, USA), pp. 91–102, ACM, 2001.
42. B. Zadrozny, J. Langford, and N. Abe, "Cost-sensitive learning by cost-proportionate example weighting," in *Proceedings of the Third IEEE International Conference on Data Mining* (Melbourne, FL, USA), IEEE Computer Society, pp. 435–442, 2003.
43. C. Elkan, "The foundations of cost-sensitive learning," in *Proceedings of the Seventeenth International Conference on Machine Learning* (Stanford, CA, USA), pp. 239–246, Morgan Kaufmann, 2001.
44. W. Fan, S. Stolfo, J. Zhang, and P. Chan, "Adacost: Misclassification cost-sensitive boosting," in *Proceedings of the Sixteenth International Conference on Machine Learning* (Bled, Slovenia), pp. 99–105, Morgan Kaufmann, 1999.
45. M. Joshi, V. Kumar, and R. C. Agarwal, "Evaluating boosting algorithms to classify rare cases: Comparison and improvements," in *First IEEE International Conference on Data Mining* (San Jose, CA, USA), IEEE Computer Society, pp. 257–264, 2001.
46. N. Chawla, A. Lazarevic, L. Hall, and K. Bowyer, "SMOTEBoost: Improving prediction of the minority class in boosting," in *Proceedings of Principles of Knowledge Discovery in Databases* (Cavtat-Dubrovnik, Croatia), pp. 107–119, Springer, 2003.
47. N. Japkowicz, C. Myers, and M. Gluck, "A novelty detection approach to classification," in *Proceedings of the Fourteenth Joint Conference on Artificial Intelligence* (Montreal, Quebec, Canada), pp. 518–523, Morgan Kaufmann, 1995.
48. B. Raskutti and A. Kowalczyk, "Extreme re-balancing for SVMs: A case study," in *ICML Workshop on Learning from Imbalanced Data Sets II* (Washington, DC), 2003.
49. W. Cohen, "Fast effective rule induction," in *Proceedings of the Twelfth International Conference on Machine Learning* (Tahoe City, CA, USA), pp. 115–123, Morgan Kaufmann, 1995.
50. M. Kubat, R. Holte, and S. Matwin, "Learning when negative examples abound," in *Machine Learning: ECML-97, Lecture Notes in Artificial Intelligence*, vol. 1224, pp. 146–153, Springer, 1997.
51. L. Breiman, J. Friedman, R. Olshen, and C. Stone, *Classification and Regressions Trees*. Boca Raton, FL: Chapman and Hall/CRC Press, 1984.

3

IMBALANCED DATASETS: FROM SAMPLING TO CLASSIFIERS

T. Ryan Hoens and Nitesh V. Chawla

Department of Computer Science and Engineering, The University of Notre Dame, Notre Dame, IN, USA

Abstract: Classification is one of the most fundamental tasks in the machine learning and data-mining communities. One of the most common challenges faced when trying to perform classification is the class imbalance problem. A dataset is considered imbalanced if the class of interest (positive or minority class) is relatively rare as compared to the other classes (negative or majority classes). As a result, the classifier can be heavily biased toward the majority class. A number of sampling approaches, ranging from under-sampling to over-sampling, have been developed to solve the problem of class imbalance. One challenge with sampling strategies is deciding how much to sample, which is obviously conditioned on the sampling strategy that is deployed. While a wrapper approach may be used to discover the sampling strategy and amounts, it can quickly become computationally prohibitive. To that end, recent research has also focused on developing novel classification algorithms that are class imbalance (skew) insensitive. In this chapter, we provide an overview of the sampling strategies as well as classification algorithms developed for countering class imbalance. In addition, we consider the issues of correctly evaluating the performance of a classifier on imbalanced datasets and present a discussion on various metrics.

3.1 INTRODUCTION

A common problem faced in data mining is dealing *class imbalance*. A dataset is said to be *imbalanced* if one class (called the *majority*, or *negative* class) vastly

Imbalanced Learning: Foundations, Algorithms, and Applications, First Edition.
Edited by Haibo He and Yunqian Ma.
© 2013 The Institute of Electrical and Electronics Engineers, Inc. Published 2013 by John Wiley & Sons, Inc.

outnumbers the other (called the *minority*, or *positive* class). The class imbalance problem is when the positive class is the class of interest.

One obvious complication that arises in the class imbalance problem is the effectiveness of accuracy (and error rate) in determining the performance of a classifiers. Consider, for example, a dataset for which the majority class represents 99% of the data, and the minority class represents 1% of the data (this dataset is said to have an imbalance ratio of 99 : 1). In such cases, the naïve classifier, which always predicts the majority class, will have an accuracy of 99%. Similarly, if a dataset has an imbalance ratio of 9999 : 1, the majority classifier will have an accuracy of 99.99%.

One consequence of this limitation can be seen when considering the performance of most traditional classifiers when applied in the class imbalance problem. This is due to the fact that the majority of traditional classifiers optimize the accuracy, and therefore generate a model that is equivalent to the naïve model described previously. Obviously such a classifier, in spite of its high accuracies, is useless in most practical applications as the minority class is often the class of interest (otherwise a classifier would not be necessary, as the class of interest almost always happens). As a result, numerous methods have been developed, which overcome the class imbalance problem. Such methods fall into two general categories, namely, sampling methods and skew-insensitive classifiers.

Sampling methods (e.g., random over-sampling and random under-sampling) have become standard approaches for improving classification performance [1]. In sampling methods, the training set is altered in such a way as to create a more balanced class distribution. The resulting sampled dataset is then more amenable to traditional data-mining algorithms, which can then be used to classify the data.

Alternatively, methods have been developed to combat the class imbalance problem directly. These methods are often specifically designed to overcome the class imbalance problem by optimizing a metric other than accuracy. By optimizing a metric other than accuracy that is more suitable for the class imbalance problem, skew-insensitive classifiers are able to generate more informative models.

In this chapter, we discuss the various approaches to overcome class imbalance, as well as various metrics, which can be used to evaluate them.

3.2 SAMPLING METHODS

Sampling is a popular methodology to counter the problem of class imbalance. The goal of sampling methods is to create a dataset that has a relatively balanced class distribution, so that traditional classifiers are better able to capture the decision boundary between the majority and the minority classes. Since the sampling methods are used to make the classification of the minority class instances easier, the resulting (sampled) dataset should represent a "reasonable" approximation of the original dataset. Specifically, the resulting dataset should contain only instances that are, in some sense, similar to those in the original dataset, that is, all instances in the modified dataset should be drawn from the same (or similar) distribution to those originally in the dataset. Note that sampling methods need not create an exactly balanced distribution, merely a distribution that the traditional classifiers are better able to handle.

Two of the first sampling methods developed were random under-sampling and random over-sampling. In the random under-sampling, the majority class instances are discarded at random until a more balanced distribution is reached. Consider, for example, a dataset consisting of 10 minority class instances and 100 majority class instances. In random under-sampling, one might attempt to create a balanced class distribution by selecting 90 majority class instances at random to be removed. The resulting dataset will then consist of 20 instances: 10 (randomly remaining) majority class instances and (the original) 10 minority class instances.

Alternatively, in random over-sampling, minority class instances are copied and repeated in the dataset until a more balanced distribution is reached. Thus, if there are two minority class instances and 100 majority class instances, traditional over-sampling would copy the two minority class instances 49 times each. The resulting dataset would then consists of 200 instances: the 100 majority class instances and 100 minority class instances (i.e., 50 copies each of the two minority class instances).

While random under-sampling and random over-sampling create more balanced distributions, they both suffer from serious drawbacks. For example, in random under-sampling (potentially), vast quantities of data are discarded. In the random under-sampling example mentioned above, for instance, roughly 82% of the data (the 90 majority class instances) was discarded. This can be highly problematic, as the loss of such data can make the decision boundary between minority and majority instances harder to learn, resulting in a loss in classification performance.

Alternatively, in random over-sampling, instances are repeated (sometimes to very high degrees). Consider the random over-sampling example mentioned above, where each instance had to be replicated 49 times in order to balance out the class distribution. By copying instances in this way, one can cause drastic overfitting to occur in the classifier, making the generalization performance of the classifier exceptionally poor. The potential for overfitting is especially true as the class imbalance ratio becomes worse, and each instance must be replicated more and more often.

In order to overcome these limitations, more sophisticated sampling techniques have been developed. We now describe some of these techniques.

3.2.1 Under-Sampling Techniques

The major drawback of random under-sampling is that potentially useful information can be discarded when samples are chosen randomly. In order to combat this, various techniques have been developed, which aim to retain all useful information present in the majority class by removing *redundant noisy*, and/or *borderline* instances from the dataset. Redundant instances are considered safe to remove as they, by definition, do not add any information about the majority class. Similarly, noisy instances are the majority class instances, which are the product of randomness in the dataset, rather than being a true representation of the underlying concept. Removing borderline instances is valuable as small perturbations to a borderline instance's features may cause the decision boundary to shift incorrectly.

One of the earliest approaches is due to Kubat and Matwin [2]. In their approach, they combine Tomek Links [3] and a modified version of the condensed nearest neighbor (CNN) rule [4] to create a directed under-sampling method. The choice to combine Tomek Links and CNN is natural, as Tomek Links can be said to remove borderline and noisy instances, while CNN removes redundant instances. To see how this works in practice, let us consider how Tomek Links and CNN are defined.

Given two instances a and b, let $\delta(a, b)$ define the distance (e.g., Euclidean) between a and b. A pair of instances a and b, which belong to different classes, is said to be a Tomek Link if there is no instance c such that $\delta(a, c) < \delta(a, b)$ or $\delta(b, c) < \delta(a, c)$. In words, instances a and b define a Tomek Link if: (i) instance a's nearest neighbor is b, (ii) instance b's nearest neighbor is a, and (iii) instances a and b belong to different classes. From this definition, we see that instances that are in Tomek Links are either boundary instances or noisy instances. This is due to the fact that only boundary instances and noisy instances will have nearest neighbors, which are from the opposite class.

By considering the entire dataset, one can remove Tomek Links by searching through each instance and removing those which fit the criteria. When using CNN, on the other hand, one builds up the dataset from a small, seed group of instances. To do this, let D be the original training set of instances, and C be a set of instances containing all of the minority class examples and a randomly selected majority instance. CNN then classifies all instances in D by finding its nearest neighbor in C and adding it to C if it is misclassified. Thus, we see that redundant instances are effectively eliminated in the resulting dataset C as any instance correctly classified by its nearest neighbors is not added to the dataset.

As an alternative to Tomek Links and CNN, another method for under-sampling is called the neighborhood cleaning rule (NCR) due to Laurikkala [5]. NCR uses Wilson's edited nearest neighbor rule (ENN) to select majority class instances to remove from the dataset. In NCR, for each instance a in the dataset, its three nearest neighbors are computed. If a is a majority class instance and is misclassified by its three nearest neighbors, then a is removed from the dataset. Alternatively, if a is a minority class instance and is misclassified by its three nearest neighbors, then the majority class instances among a's neighbors are removed.

3.2.2 Over-Sampling Techniques

While random under-sampling suffered from the loss of potentially useful information, random over-sampling suffers from the problem of overfitting. Specifically, by randomly replicating instances in the dataset, the learned model might fit the training data too closely and, as a result, not generalize well to unseen instances.

In order to overcome this issue, Chawla et al. developed a method of creating synthetic instances instead of merely copying existing instances in the dataset. This techniques is known as the synthetic minority over-sampling technique (SMOTE) [6]. As mentioned, in SMOTE, the training set is altered by adding synthetically generated minority class instances, causing the class distribution to become more balanced. We say that the instances created are *synthetic*, as they are, in general, new minority instances that have been extrapolated and created out of existing minority class instances.

To create the new synthetic minority class instances, SMOTE first selects a minority class instance a at random and finds its k nearest minority class neighbors. The synthetic instance is then created by choosing one of the k nearest neighbors b at random and connecting a and b to form a line segment in the feature space. The synthetic instances are generated as a convex combination of the two chosen instances a and b.

Given SMOTE's effectiveness as an over-sampling method, it has been extended multiple times [7–9]. In Borderline-SMOTE, for instance, only borderline instances are considered to be SMOTEd, where borderline instances are defined as instances that are misclassified by a nearest neighbor classifier. Safe-Level-SMOTE, on the other hand, defines a "safe-level" for each instance, and the instances that are deemed "safe" are considered to be SMOTEd.

In addition to SMOTE, Jo and Japkowicz [10] defined an over-sampling method based on clustering. That is, instead of randomly choosing the instances to oversample, they instead first cluster all of the minority class instances using k-means clustering. They then oversample each of the clusters to have the same number of instances, and the overall dataset to be balanced. The purpose of this method is to identify the disparate regions in the feature space where minority class instances are found and to ensure that each region is equally represented with minority class instances.

In addition to cluster-based over-sampling, Japkowicz et al. [11] also developed a method called *focused resampling*. In focused resampling, only minority class instances that occur on the boundary between minority and majority class instances are over-sampled. In this way, redundant instances are reduced, and better performance can be achieved.

3.2.3 Hybrid Techniques

In addition to merely over-sampling or under-sampling the dataset, techniques have been developed, which perform a combination of both. By combining over-sampling and under-sampling, the dataset can be balanced by neither losing too much information (i.e., under-sampling too many majority class instances), nor suffering from overfitting (i.e., over-sampling too heavily).

Two examples of hybrid techniques that have been developed include SMOTE+Tomek and SMOTE+ENN [12], wherein SMOTE is used to oversample the minority class, while Tomek and ENN, respectively, are used to under-sample the majority class.

3.2.4 Ensemble-Based Methods

One popular approach toward improving performance for classification problems is to use ensembles. Ensemble methods aim to leverage the classification power of multiple base learners (learned on different subsets of the training data) to improve on the classification performance over traditional classification algorithms. Dietterich [13] provides a broad overview as to why ensemble methods often outperform a single classifier. In fact, Hansen and Salamon [14] prove that under certain constraints (the average error rate is less than 50% and the probability of misprediction of each classifier is independent of the others), the expected error rate of an instance goes to zero as the number of classifiers goes to infinity. Thus, when seeking to build multiple classifiers, it is better to ensure that the classifiers are diverse rather than highly accurate.

There are many popular methods for building diverse ensembles, including bagging [15], AdaBoost [16], Random Subspaces [17], and Random Forests [18]. While each of these ensemble methods can be applied to datasets that have undergone sampling, in general, this is not optimal as it ignores the power of combining the ensemble generation method and sampling to create a more structured approach. As a result, many ensemble methods have been combined with sampling strategies to create ensemble methods that are more suitable for dealing with class imbalance.

AdaBoost is one of the most popular ensemble methods in the machine learning community due, in part, to its attractive theoretical guarantees [16]. As a result of its popularity, AdaBoost has undergone extensive empirical research [13, 19]. Recall that in AdaBoost, base classifier L is learned on a subset S_L of the training data D, where each instance in S_L is probabilistically selected on the basis of its weight in D. After training each classifier, each instance's weight is adaptively updated on the basis of the performance of the ensemble on the instance. By giving more weight to misclassified instances, the ensemble is able to focus on instances that are difficult to learn.

SMOTEBoost is one example of combining sampling methods with AdaBoost to create an ensemble that explicitly aims to overcome the class imbalance [20]. In SMOTEBoost, in addition to updating instance weights during each boosting iteration, SMOTE is also applied to misclassified minority class examples. Thus, in addition to emphasizing minority instances by giving higher weights, misclassified minority instances are also emphasized by the addition of (similar) synthetic examples. Similar to SMOTEBoost, Guo and Viktor [21] develop another extension for boosting called *DataBoost-IM*, which identifies hard instances (both minority and majority) in order to generate similar synthetic examples and then reweights the instances to prevent a bias toward the majority class.

An alternative to AdaBoost is Bagging, another ensemble method that has been adapted to use sampling. Radivojac et al. [22] combine bagging with over-sampling techniques in the bioinformatics domain. Liu et al. propose two methods, EasyEnsemble and BalanceCascade [23], that generate training sets by choosing an equal number of majority and minority class instances from the training set. Hido and Kashima [24] introduce a variant of bagging, "Roughly Balanced Bagging" (RB bagging), that alters bagging to emphasize the minority class.

Finally, Hoens and Chawla [25] propose a method called *RSM+SMOTE* that involves combining random subspaces with SMOTE. Specifically, they note that SMOTE depends on the nearest neighbors of an instance to generate synthetic instances. Therefore, by choosing different sets of features to apply SMOTE in (and thereby altering the nearest neighbor calculation used by SMOTE to create the synthetic instance), the resulting training data for each base learner will have different biases, promoting a more diverse—and therefore effective—ensemble.

3.2.5 Drawbacks of Sampling Techniques

One major drawback of sampling techniques is that one needs to determine how much sampling to apply. An over-sampling level must be chosen so as to promote the minority class, while avoiding overfitting to the given data. Similarly, an under-sampling level must be chosen so as to retain as much information about the majority class as possible, while promoting a balanced class distribution.

In general, wrapper methods are used to solve this problem. In wrapper methods, the training data is split into a training set and a validation set. For a variety of sampling levels, classifiers are learned on the training set. The performance of each of the learned models is then tested against the validation set. The sampling method that provides the best performance is then used to sample the entire dataset.

For hybrid techniques, such wrappers become very complicated, as instead of having to optimize a single over- (or under-) sampling level, one has to optimize a combination of over- and under-sampling levels. As demonstrated by Cieslak et al. [26], such wrapper techniques are often less effective at combating class imbalance than ensembles built of skew- insensitive classifiers. As a result, we now turn our focus to skew-insensitive classifiers.

3.3 SKEW-INSENSITIVE CLASSIFIERS FOR CLASS IMBALANCE

While sampling methods—and ensemble methods based on sampling methods—have become the de facto standard for learning in datasets that exhibit class imbalance, methods have also been developed that aim to directly combat class imbalance without the need for sampling. These methods come mainly from the cost-sensitive learning community; however, classifiers that deal with imbalance are not necessarily cost-sensitive learners.

3.3.1 Cost-Sensitive Learning

In cost-sensitive learning instead of each instance being either correctly or incorrectly classified, each class (or instance) is given a *misclassification cost*. Thus, instead of trying to optimize the accuracy, the problem is then to minimize the total misclassification cost. Many traditional methods are easily extensible with this goal in mind. Support vector machines (SVMs), for instance, can be easily modified to follow the cost-sensitive framework.

To see how this is done, consider the problem of learning an SVM. SVMs attempt to learn a weight vector **w**that satisfies the following optimization problem:

$$\min_{w,\xi,b} \left\{ \frac{1}{2}||w||^2 + C \sum_{i=1}^{n} \xi_i \right\} \tag{3.1}$$

subject to (for all $i \in \{1, \ldots, n\}$):

$$y_i(\mathbf{w} \cdot \mathbf{x}_i - b) \geq 1 - \xi_i \tag{3.2}$$

$$\xi_i \geq 0 \tag{3.3}$$

where $\mathbf{x}_i$ denotes the instance i, y_i denotes the class of instance i, ξ_i denotes the "slack," for instance, i (i.e., how badly misclassified, if at all, instance i is by **w**), and C determines the trade-off between training error and model complexity.

In order to make SVMs cost sensitive, the objective function is modified resulting in:

$$\min_{w,\xi,b} \left\{ \frac{1}{2}||w||^2 + C \sum_{i=1}^{n} c_i \xi_i \right\}, \tag{3.4}$$

where c_i is the misclassification cost, for instance i.

In addition to modifying traditional learning algorithms, cost-sensitive ensemble methods have also been developed. AdaCost [27] and cost-sensitive boosting (CSB) [28] are two extensions of AdaBoost, which incorporate the misclassification cost of an instance in order to provide more accurate instance weights and not just weights derived from misclassification error as done by AdaBoost.

3.3.2 Skew-Insensitive Learners

While cost-sensitive learning is a popular way of extending classifiers for the use on the class imbalance problem, some classifiers allow for easier extensions that directly combat the problem of class imbalance.

Naïve Bayes, for instance, is trivially skew insensitive. This is due to the fact that it makes predictions $p(y|\mathbf{x}_i)$ by first computing $p(\mathbf{x}_i|y)$ and $p(y)$ from the training data. Bayes rule is then applied to obtain a classification, for instance, $\mathbf{x}_i$ as: $p(y|\mathbf{x}_i) = p(\mathbf{x}_i|y) \cdot p(y)$. Since $p(\mathbf{x}_i|y)$ is difficult to calculate in general, a simplifying assumption is made (the "naïve" in "naïve Bayes"), namely, each of the features is assumed independent. With this assumption, we can compute $p(\mathbf{x}_i|y) = \prod_{\forall j,k} p(\mathbf{x}_{ij} = \mathbf{x}_{ijk}|y)$, where $\mathbf{x}_{ij}$ denotes feature j, for instance, i, and $\mathbf{x}_{ijk}$ denotes the kth possible feature value for feature j. Therefore, naïve Bayes is simply skew insensitive as predictions are calibrated by $p(y)$ or the prior probability of class y.

Another classifier that has recently been made skew insensitive are decision trees. Hellinger distance decision trees (HDDTs) [29] are strongly skew insensitive, using an adaptation of the Hellinger distance as a decision tree splitting criterion. They mitigate the need for sampling.

For the sake of clarity, we present the basic decision tree-building algorithm. The algorithm (Algorithm $Build\_Tree$) differs from the traditional C4.5 [30] algorithm in two important facets, both motivated by the research of Provost and Domingos [31]. First, when building the decision tree, $Build\_Tree$ does not consider pruning or collapsing. Second, when classifying an instance, Laplace smoothing is applied. These choices are because of empirical results, demonstrating that a full tree with Laplace smoothing outperforms all other configurations [31], which are particularly relevant for imbalanced datasets. When C4.5 decision trees are built in this way (i.e., without pruning, without collapsing, and with Laplace smoothing), they are called *C4.4 decision trees* [31].

Algorithm $Build\_Tree$

Require: Training set T, Cut-off size C
 1: **if** $|T| < C$ **then**
 2:     **return**
 3: **end if**
 4: **for** each feature f of T **do**
 5:     $H_f \leftarrow Calc\_Criterion\_Value(T, f)$
 6: **end for**
 7: $b \leftarrow \max(H)$
 8: **for** each value $\mathbf{v}$ of b **do**
 9:     $Build\_Tree(T_{x_b = v}, C)$
10: **end for**

An important thing to note is that $Build\_Tree$ is only defined for nominal features. For continuous features, a slight variation to $Build\_Tree$ is used, where $Calc\_Criterion\_Value$ sorts the instances by the feature value, finds all meaningful splits, calculates the binary criterion value at each split, and returns the highest distance; this is identical to the procedure used in C4.5.

The important function to consider when building a decision tree is $Calc\_Criterion\_Value$. In C4.5, this function is gain ratio, which is a measure of purity based on entropy [30], while in HDDT, this function is Hellinger distance. We now describe the Hellinger distance as a splitting criterion.

Hellinger distance is a distance metric between probability distributions used by Cieslak and Chawla [29] to create HDDTs. It was chosen as a splitting criterion for the binary class imbalance problem because of its property of skew insensitivity. Hellinger distance is defined as a splitting criterion as [29]:

$$d_H(X_+, X_-) = \sqrt{\sum_{j=1}^{p} \left(\sqrt{\frac{|X_{+j}|}{|X_+|}} - \sqrt{\frac{|X_{-j}|}{|X_-|}} \right)^2} \tag{3.5}$$

where X_+ is the set of all positive examples, X_- is the set of all negative examples, and X_{+j} (X_{-j}) is the set of positive (negative) examples with the jth value of the relevant feature.

3.4 EVALUATION METRICS

One common method for determining the performance of a classifier is through the use of a confusion matrix (Fig. 3.1). In a confusion matrix, TN is the number of negative instances correctly classified (True Negatives), FP is the number of negative instances incorrectly classified as positive (False Positive), FN is the number of positive instances incorrectly classified as negative (False Negatives), and TP is the number of positive instances correctly classified as positive (True Positives).

From the confusion matrix, many standard evaluation metrics can be defined. Traditionally, the most often used metric is accuracy (or its complement, the error rate):

$$\text{accuracy} = \frac{TP + TN}{TP + FP + TN + FN} \tag{3.6}$$

As mentioned previously, however, accuracy is inappropriate when data is imbalanced. This is seen in our previous example, where the majority class may

	Predicted negative	Predicted positive
Actual negative	TN	FP
Actual positive	FN	TP

Figure 3.1 Confusion matrix.

make 99% of the dataset, while only 1% is made up of minority class instances. In such a case, it is trivial to obtain an accuracy of 99% if one merely predicts the majority class. This accuracy is obtained by simply predicting all instances as majority class. While the 99% accuracy seems high, it is obviously completely discounting the performance of the classifier on the class that matters (positive class). Thus, accuracy can be misleading in imbalanced datasets as there are two different types of errors (false positives and false negatives), and each of them carries different costs.

We now present a discussion on a number of alternate metrics used for evaluating the performance of classifiers on imbalanced datasets.

3.4.1 Balanced Accuracy

While accuracy (and error rate) is not an effective method of evaluating the performance of classifiers, one common alternative is *balanced accuracy*. Balanced accuracy differs from accuracy in that instead of computing *accuracy*, one computes

$$\text{BalancedAccuracy} = \frac{\text{TP}}{2(\text{TP} + \text{FN}))} + \frac{\text{TN}}{2(\text{TN} + \text{FP}))} \qquad (3.7)$$

That is, one computes the average of the percentage of positive class instances correctly classified and the percentage of negative class instances correctly classified. By giving equal weight to these relative proportions, we see that the previous problem of the naïve classifier obtaining very good performance has been eliminated.

To see this, consider the balanced accuracy of the naïve classifier on a dataset consisting of 99% majority class instances and 1% minority class instances. We know that the accuracy of the naïve classifier is 99%. The balanced accuracy, on the other hand, is: $99/(2(99 + 0)) + 0/(2(1 + 0)) = 0.5 + 0 = 0.5$. A performance estimate of 0.5 is a much more valid assessment of the naïve classifier.

3.4.2 ROC Curves

The receiver operating characteristic (ROC) curve is a standard technique for evaluating classifiers on datasets that exhibit class imbalance. ROC curves achieve this skew insensitivity by summarizing the performance of classifiers over a range of true positive rates (TPRs) and false positive rates (FPRs) [32]. By evaluating the models at various error rates, ROC curves are able to determine what proportion of instances will be correctly classified for a given FPR.

In Figure 3.2, we see an example of an ROC curve. In Figure 3.2, the X-axis represents the FPR (FPR = FP/(TN + FP)), and the Y-axis represents the TPR (TPR = TP/(TP + FN)). For any given problem, the ideal classifier would have

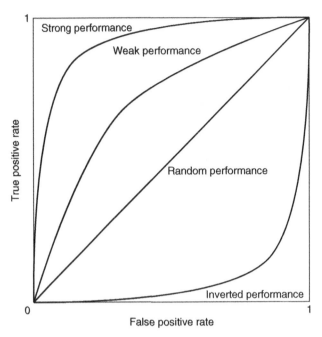

Figure 3.2   Examples of ROC curves. Each curve represents the performance of a different classifier on a dataset.

a point in the ROC space at $(0, 1)$, that is, all positive instances are correctly classified, and no negative instances are misclassified. Alternatively, the classifier that misclassifies all instances would have a single point at $(1, 0)$.

While $(0, 1)$ represents the ideal classifier and $(1, 0)$ represents its complement, in ROC space, the line $y = x$ represents a random classifier, that is, a classifier that applies a random prediction to each instance. This gives a trivial lower bound in ROC space for any classifier. An ROC curve is said to "dominate" another ROC curve if, for each FPR, it offers a higher TPR. By analyzing ROC curves, one can determine the best classifier for a specific FPR by selecting the classifier with the best corresponding TPR.

In order to generate an ROC curve, each point is generated by moving the decision boundary for classification. That is, points nearer to the left in ROC space are the result of requiring a higher threshold for classifying an instance as positive instance. This property of ROC curves allows practitioners to choose the decision threshold that gives the best TPR for an acceptable FPR (Neyman–Pearson method) [33].

The ROC convex hull can also provide a robust method for identifying potentially optimal classifiers [34]. Given a set of ROC curves, the ROC convex hull is generated by selecting only the best point for a given FPR. This is advantageous, since, if a line passes through a point on the convex hull, then there is no other line with the same slope passing through another point with a larger TPR intercept. Thus, the classifier at that point is optimal under any distribution assumptions along with the slope [34].

While ROC curves provide a visual method for determining the effectiveness of a classifier, the area under the ROC curve (AUROC) has become the de facto standard metric for evaluating classifiers under imbalance [35]. This is due to the fact that it is both independent of the selected threshold and prior probabilities, as well as offering a single number to compare classifiers. One of the main benefits of AUROC is that it can be considered as measuring how often a random positive class instance is ranked above a random negative class instance when sorted by their classification probabilities.

One way of computing AUROC is, given n_0 points of class 0, n_1 points of class 1, and S_0 as the sum of ranks of class 0 examples [36]:

$$\text{AUROC} = \frac{2S_0 - n_0(n_0 + 1)}{2n_0 n_1} \tag{3.8}$$

3.4.3 Precision and Recall

Alternatives to AUROC are precision and recall. Precision and recall can be computed from the confusion matrix (Fig. 3.1) as [37]:

$$\text{precision} = \frac{\text{TP}}{\text{TP} + \text{FP}} \tag{3.9}$$

$$\text{recall} = \frac{\text{TP}}{\text{TP} + \text{FN}} \tag{3.10}$$

From the equations, we see that precision measures how often an instance that was predicted as positive that is actually positive, while recall measures how often a positive class instance in the dataset was predicted as a positive class instance by the classifier.

In imbalanced datasets, the goal is to improve recall without hurting precision. These goals, however, are often conflicting, since in order to increase the TP for the minority class, the number of FP is also often increased, resulting in reduced precision.

In order to obtain a more accurate understanding of the trade-offs between precision and recall, one can use precision–recall (PR) curves. PR curves are similar to ROC curves in that they provide a graphical representation of the performance of classifiers. While the X-axis of ROC curves is FPR and the Y-axis is TPR, in PR curves, the X-axis is recall and the Y-axis is precision. Precision–recall curves are therefore similar to ROC curves as recall is the same as FPR; however, the Y-axes are different. While TPR measures the fraction of positive examples that are correctly classified, precision measures the fraction of examples that are classified as positive that are actually positive.

Similar to ROC curves are being compared on the basis of AUROC, PR curves are also compared on the basis of AUPR. This practice has become more common, as recent research suggests that PR curves (and AUPR) are a better discriminator of performance than their ROC (and AUROC) counterparts [38].

3.4.4 F_β-Measure

A final common metric is the F_β-measure. F_β-measure is a family of metrics that attempts to measure the trade-offs between precision and recall by outputting a single value that reflects the *goodness* of a classifier in the presence of rare classes. While ROC curves represent the trade-off between different TPRs and FPRs, F_β-measure represents the trade-off among different values of TP, FP, and FN [37].

The general equation for F_β-measure is:

$$F_\beta = (1 + \beta^2) \cdot \frac{\text{precision} \cdot \text{recall}}{(\beta^2 \cdot \text{precision}) + \text{recall}}, \tag{3.11}$$

where β represents the relative importance of precision and recall. Traditionally, when β is not specified, the F_1-measure is assumed.

In spite of its (relatively) useful properties for imbalance, F_β-measure is not commonly used when comparing classifiers, as AUROC and AUPR provide more robust and better performance estimates.

3.5 DISCUSSION

In this chapter, we covered various strategies for learning in imbalanced environments. Specifically, we discussed various sampling strategies and skew- insensitive classifiers.

One key observation when attempting to choose between a sampling method and a skew-insensitive classifier is that while sampling methods are a widely applied standard, they require the tuning of parameters to select the proper sampling level for a given dataset. In general, this is a difficult optimization problem and may prove impractical in practice depending on the size of the dataset and level of imbalance. In such cases, skew-insensitive classifiers (and ensembles built of skew-insensitive classifiers) can provide a reasonable alternative that provides similar (and often better) performance as that of the sampling methods.

When attempting to evaluate the performance of the aforementioned models, we learned that accuracy is not a valuable evaluation metric when learning in imbalanced environments. The lack of utility of accuracy (and error rate) is due to the fact that they overemphasize the performance of the majority class at the detriment to the considerations of the performance of the minority class. In order to overcome this issue, we presented multiple alternative evaluation metrics. The most commonly used alternatives discussed were AUROC and AUPR.

REFERENCES

1. N. V. Chawla, D. A. Cieslak, L. O. Hall, and A. Joshi, "Automatically countering imbalance and its empirical relationship to cost," *Data Mining and Knowledge Discovery*, vol. 17, no. 2, pp. 225–252, 2008.

2. M. Kubat and S. Matwin, "Addressing the curse of imbalanced training sets: One-sided selection," in *Machine Learning-International Workshop then Conference*, (Nashville, TN, USA), pp. 179–186. Morgan Kaufmann, 1997.

3. I. Tomek, "An experiment with the edited nearest-neighbor rule," *IEEE Transactions on Systems, Man, and Cybernetics Part C*, vol. 6, no. 6, pp. 448–452, 1976.

4. P. E. Hart, N. J. Nilsson, and B. Raphael, "A formal basis for the heuristic determination of minimum cost paths," *IEEE Transactions on Systems Science and Cybernetics*, vol. 4, no. 2, pp. 100–107, 1968.

5. J. Laurikkala, "Improving identification of difficult small classes by balancing class distribution," *Artificial Intelligence in Medicine*, (Cascais, Portugal), pp. 63–66, Springer-Verlag, vol. 2101, 2001.

6. N. V. Chawla, K. W. Bowyer, L. O. Hall, and W. P. Kegelmeyer, "SMOTE: Synthetic minority over-sampling technique," *Journal of Artificial Intelligence Research*, vol. 16, pp. 321–357, 2002.

7. H. Han, W. Y. Wang, and B. H. Mao, "Borderline-smote: A new over-sampling method in imbalanced data sets learning," in *Advances in Intelligent Computing*, (Hefei, China), vol. 3644, pp. 878–887, Springer-Verlag, 2005.

8. C. Bunkhumpornpat, K. Sinapiromsaran, and C. Lursinsap, "Safe-level-smote: Safe-level-synthetic minority over-sampling technique for handling the class imbalanced problem," in *Pacific-Asia Conference on Advances in Knowledge Discovery and Data Mining*, (Bangkok, Thailand), pp. 475–482, Springer-Verlag, 2009.

9. C. Bunkhumpornpat, K. Sinapiromsaran, and C. Lursinsap, "DBSMOTE: Density-based synthetic minority over-sampling technique," *Applied Intelligence*, vol. 36, pp. 1–21, 2011.

10. T. Jo and N. Japkowicz, "Class imbalances versus small disjuncts", *ACM SIGKDD Explorations Newsletter*, vol. 6, no. 1, pp. 40–49, 2004.

11. N. Japkowicz "Learning from imbalanced data sets: A comparison of various strategies," in *AAAI Workshop on Learning from Imbalanced Data Sets*, (Austin, Texas), vol. 68, AAAI Press, 2000.

12. G. E. Batista, R. C. Prati, and M. C. Monard, "A study of the behavior of several methods for balancing machine learning training data," *ACM SIGKDD Explorations Newsletter*, vol. 6, no. 1, pp. 20–29, 2004.

13. T. G. Dietterich, "Ensemble methods in machine learning," *Lecture Notes in Computer Science*, vol. 1857, pp. 1–15, 2000.

14. L. K. Hansen and P. Salamon, "Neural network ensembles," *IEEE Transactions on Pattern Analysis and Machine Intelligence*, vol. 12, no. 10, pp. 993–1001, 1990.

15. L. Breiman, "Bagging predictors," *Machine Learning*, vol. 24, no. 2, pp. 123–140, 1996.

16. Y. Freund and R. Schapire. "Experiments with a new boosting algorithm," in *Thirteenth International Conference on Machine Learning*, (Bari, Italy), pp. 148–156, Morgan Kaufmann, 1996.

17. T. K. Ho, "The random subspace method for constructing decision forests," *IEEE Transactions on Pattern Analysis and Machine Intelligence*, vol. 20, no. 8, pp. 832–844, 1998.

18. L. Breiman, "Random forests," *Machine Learning*, vol. 45, no. 1, pp. 5–32, 2001.

19. E. Bauer and R. Kohavi, "An empirical comparison of voting classification algorithms: Bagging, boosting, and variants," *Machine Learning*, vol. 36, no. 1, pp. 105–139, 1999.

20. N. V. Chawla, A. Lazarevic, L. O. Hall, and K. W. Bowyer. "Smoteboost: Improving prediction of the minority class in boosting," in *Proceedings of the Principles of Knowledge Discovery in Databases, PKDD-2003*, (Cavtat-Dubrovnik, Croatia), vol. 2838, pp.107–119, Springer-Verlag, 2003.

21. H. Guo and H. L. Viktor, "Learning from imbalanced data sets with boosting and data generation: The Databoost-IM approach," *SIGKDD Explorations Newsletter*, vol. 6, no. 1, pp. 30–39, 2004.

22. P. Radivojac, N. V. Chawla, A. K. Dunker, and Z. Obradovic, "Classification and knowledge discovery in protein databases," *Journal of Biomedical Informatics*, vol. 37, no. 4, pp. 224–239, 2004.

23. X. Y. Liu, J. Wu, and Z. H. Zhou. "Exploratory under-sampling for class-imbalance learning", in *ICDM '06: Proceedings of the Sixth International Conference on Data Mining*, pp. 965–969 Washington, DC: IEEE Computer Society, 2006.

24. S Hido and H. Kashima, "Roughly balanced bagging for imbalanced data," in *SDM*, pp. 143–152. SIAM, 2008.

25. T. Hoens and N. Chawla, "Pacific-Asia Conference on Advances in Knowledge Discovery and Data Mining," in *PAKDD*, (Hyderabad, India), vol. 6119, pp. 488–499, Springer-Verlag, 2010.

26. D. A. Cieslak, T. R. Hoens, N. V. Chawla, and W. P. Kegelmeyer, "Hellinger distance decision trees are robust and skew-insensitive," *Data Mining and Knowledge Discovery*, (Hingham, MA, USA), vol. 24, no. 1, pp. 136–158, Kluwer Academic Publishers, 2012.

27. W. Fan, S. J. Stolfo, J. Zhang, and P. K. Chan. "Adacost: Misclassification cost-sensitive boosting," in *Machine Learning-International Workshop then Conference*, (Bled, Slovenia), pp. 97–105, Morgan Kaufmann, 1999.

28. K. M. Ting. "A comparative study of cost-sensitive boosting algorithms," in *Proceedings of the 17th International Conference on Machine Learning*, 2000.

29. D. A. Cieslak and N. V. Chawla. "Learning decision trees for unbalanced data," in *European Conference on Machine Learning (ECML)*, (Antwerp, Belgium), vol. 5211, pp. 241–256, Springer-Verlag, 2008.

30. J. R. Quinlan. *C4. 5: Programs for Machine Learning*. San Mateo, CA: Morgan Kaufman Publishers, Inc., 1993.

31. F. Provost and P. Domingos, "Tree induction for probability-based ranking," *Machine Learning*, vol. 52, no. 3, pp. 199–215, 2003.

32. J. A. Swets, "Measuring the accuracy of diagnostic systems," *Science*, vol. 240, no. 4857, pp. 1285–, 1988.

33. J. P. Egan. *Signal Detection Theory and ROC Analysis*. New York: Academic Press, 1975.

34. F. Provost and T. Fawcett, "Robust classification for imprecise environments," *Machine Learning*, vol. 42, no. 3, pp. 203–231, 2001.

35. A. P. Bradley, "The use of the area under the roc curve in the evaluation of machine learning algorithms," *Pattern Recognition*, vol. 30, no. 7, pp. 1145–1159, 1997.

36. D. J. Hand and R. J. Till, "A simple generalisation of the area under the roc curve for multiple class classification problems," *Machine Learning*, vol. 45, no. 2, pp. 171–186, 2001.

37. M. Buckland and F. Gey, "The relationship between recall and precision," *Journal of the American Society for Information Science*, vol. 45, no. 1, pp. 12–19, 1994.

38. J. Davis and M. Goadrich, "The relationship between precision-recall and roc curves," in *Proceedings of the Twentythird International Conference on Machine Learning*, pp. 233–240. ACM, 2006.

4

ENSEMBLE METHODS FOR CLASS IMBALANCE LEARNING

Xu-Ying Liu

School of Computer Science and Engineering, Southeast University, Nanjing, China

Zhi-Hua Zhou

National Key Laboratory for Novel Software Technology, Nanjing University, Nanjing, China

Abstract: Ensemble learning is an important paradigm in machine learning, which uses a set of classifiers to make predictions. The generalization ability of an ensemble is generally much stronger than that of individual ensemble members. Ensemble learning is widely exploited in the literature of class imbalance learning. This chapter introduces ensemble learning and gives an overview of ensemble methods for class imbalance learning.

4.1 INTRODUCTION

Ensemble methods use a set of classifiers to make predictions. The generalization ability of an ensemble is usually much stronger than that of the individual ensemble members. Ensemble learning is one of the main learning paradigms in machine learning and has achieved great success in almost everywhere learning methods are applicable, such as object detection, face recognition, recommending systems, medical diagnosis, text categorization, and so on. For example, an ensemble architecture was proposed by Huang et al. [1] for pose-invariant face recognition. It trains a set of view-specific neural networks (NNs) and then uses sophisticated combination rules to form an ensemble. Conventional methods train a single classifier and require pose information as input, while the ensemble

Imbalanced Learning: Foundations, Algorithms, and Applications, First Edition.
Edited by Haibo He and Yunqian Ma.
© 2013 The Institute of Electrical and Electronics Engineers, Inc. Published 2013 by John Wiley & Sons, Inc.

architecture does not need it and can further output pose information in addition to the face recognition result, and its classification accuracy is higher than that of the conventional methods supplied with precise pose information. Moreover, it is not a surprise that in various real-world data-mining competitions, such as KDD-Cup[1] and Netflix Prize,[2] almost all the top algorithms exploited ensemble methods in recent years.

In class imbalance learning (CIL), ensemble methods are broadly used to further improve the existing methods or help design brand new ones. A famous example is the ensemble method designed by Viola and Jones [2, 3] for face detection. Face detection requires to indicate which parts of an image contain a face in real-time. A typical image has about 50,000 sub-windows to represent different scales and locations [2], and each one of them should be determined whether it contains a face or not. Typically, there are only a few dozen faces among these sub-windows in an image. Furthermore, there are often more non-face images than images containing any face. Thus, non-face sub-windows could be 10^4 times more than sub-windows containing any face. Viola and Jones [2, 3] designed a boosting-based ensemble method to deal with the severe class imbalance problem. This method, together with a cascade-style learning structure, is able to achieve very high detection rate while keeping very low false positive rate. This face detector is recognized as one of the breakthroughs in the past decades. Besides, ensemble methods have been used to improve over-sampling [4] and under-sampling [5, 6], and a number of boosting-based methods have been developed to handle class-imbalanced data [2, 4, 7, 8].

We introduce the notations used in this chapter. By default, we talk about binary classification problems. Let $D = \{(\mathbf{x}_i, y_i)\}_{i=1}^n$ be the training set, with $y \in \{-1, 1\}$. The class with $y = 1$ is the positive class with n_+ examples, and suppose it is the minority class; the class with $y = -1$ is the negative class with n_- examples, and suppose it is the majority class. So we have $n_+ < n_-$, and the level of imbalance is $r = n_-/n_+$. The subset of training data containing all the minority class examples is $\mathcal{P}$, and the subset containing all the majority class examples is $\mathcal{N}$. Assume that data is independent and identically sampled from distribution $\mathcal{D}$ on $\mathcal{X} \times \mathcal{Y}$, where $\mathcal{X}$ is the input space and $\mathcal{Y}$ is the output space. A learning algorithm $\mathfrak{L}$ trains a classifier $h : \mathcal{X} \to \mathcal{Y}$.

4.2 ENSEMBLE METHODS

The most central concept of machine learning is generalization ability, which indicates how well the unseen data could be predicted by the learner trained

---

[1]KDD-Cup is the most famous data-mining competition that covers various real-world applications, such as network intrusion, bioinformatics, and customer relationship management. For further details, refer to http://www.sigkdd.org/kddcup/
[2]Netflix is an online digital video disk (DVD) rental service. Netflix Prize is a data-mining competition held every year since 2007 to help improve the accuracy of movie recommendation for users. For further details, refer to http://www.netflixprize.com/

from the training data. Many efforts have been devoted to design methods that can generate learners with strong generalization ability. Ensemble learning is one of the most successful paradigms. Different from ordinary machine learning methods (which usually generate one single learner), ensemble methods train a set of *base learners* from training data to make predictions with each one of them, and then combine these predictions to give the final decision.

The most amazing part of the ensemble is that it can boost learners with slightly better performance than random guess into learners with strong generalization ability. Thus, the "base learners" are often referred as *weak learners*. This also indicates that in ensemble methods, the "base learners" can have weak generalization ability. Actually, most learning algorithms, such as decision trees, NN, or other machine learning methods, can be invoked to train "base learners," and ensemble methods can boost the performance.

According to how the base learners are generated, ensemble methods can be roughly categorized into two paradigms: parallel ensemble methods and sequential ensemble methods. Parallel ensemble methods generate base learners in parallel, with Bagging [9] as a representative. Sequential ensemble methods generate base learners in sequential, where a former base learner has influence on the generation of subsequent learners, with AdaBoost [10] as a representative. We will briefly introduce Bagging and AdaBoost in Sections 4.2.1 and 4.2.2. After generating base learners, rather than trying to use the best individual learner, ensemble methods combine them with a certain combination method. There are several popular combination methods, such as averaging, voting, and stacking [11–13].

Generally speaking, to get a good ensemble, base learners should be as more accurate as possible and more diverse as possible, which is formally shown by Krogh and Vedelsby [14], and emphasized and used by many people. The diversity of base learners can be obtained in different ways, such as sampling the training data, manipulating the attributes, manipulating the outputs, injecting randomness into learning process, or even using multiple mechanisms simultaneously. For comprehensive introduction of ensemble learning, please refer to [15].

Algorithm The Bagging algorithm for classification

Input: Data set $D = \{(\mathbf{x}_i, y_i)\}_{i=1}^{n}$
        Base learning algorithm $\mathfrak{L}$
        The number of iterations T
1: **for** $t = 1$ to T **do**
2:    $h_t = \mathfrak{L}(D, \mathcal{D}_{bs})$ /* $\mathcal{D}_{bs}$ is the bootstrap distribution */
3: **end for**
Output: $H(\mathbf{x}) = \max\limits_{y} \sum_{t=1}^{T} \mathrm{I}(h_t(\mathbf{x}) = y)$
                /* I(x) = 1 if x is true, and 0 otherwise */

4.2.1 Bagging

The basic motivation of parallel ensemble methods is that error can be reduced dramatically by combining independent base learners. Since it is difficult to get totally independent base learners from the same set of training data in practice, base learners with less dependency can be obtained by injecting randomness in the learning process.

Bagging [9] is a representative of parallel ensemble method. The name is short for "Bootstrap AGGregatING," indicating two principal techniques of Bagging, that is, bootstrap and aggregation. Bagging adopts bootstrap sampling [16] to inject randomness into training data to generate base learners with less dependency and more diversity. In detail, given the original training set of n examples $D = \{(\mathbf{x}_i, y_i)\}_{i=1}^{n}$, a sample of n training examples will be generated by sampling with replacement. Thus, some original training data appear several times. By applying the sampling with replacement for T times, T subsets of n training examples are obtained, which can be used to train T base learners. Bagging uses the most popular combination method voting for classification and averaging for regression to form an ensemble. The algorithm is shown previously. It is worth noting that, because the base learners are independent, the computational cost of Bagging can be further reduced by training base learners in parallel.

It is quite remarkable that, although the training data for different base learners have diversity by injecting randomness, this does not necessarily lead to diverse base learners. Some learning methods are very stable; they are insensitive to perturbation on training samples, such as decision stumps and k-nearest neighbor. The base learners generated by these methods are quite similar to each other; so combining them will not improve generalization ability [15]. Therefore, Bagging should be used with unstable learners, such as decision trees. Its generalization ability is improved by reducing the variance through the smoothing effect [9]. And generally, the more unstable the base learners, the larger the performance improvement.

4.2.2 AdaBoost

Boosting-style ensemble methods can boost weak learners into strong learners. Weak learners are slightly better than random guess, while strong learners can predict very accurately. The research of boosting derived from the work on an interesting theoretical question posed by Kearns and Valiant [17]: whether the weakly learnable problems equal the strongly learnable problems. If the answer to the question is positive, the strongly learnable problems can be solved by "boosting" a series of weak learners. Obviously, this question is of fundamental importance to machine learning. Schapire [18] gave a constructive proof to this question and this led to the first boosting algorithm. The basic idea of boosting is to correct the mistakes of previous base learners. Specifically, suppose that the first base learner h_1 trained from the original training data is a weak learner. Boosting will derive a new distribution $\mathcal{D}'$ from the true distribution $\mathcal{D}$ to focus

---

Algorithm The AdaBoost algorithm

Input: Data set $D = \{(\mathbf{x}_i, y_i)\}_{i=1}^n$
Base learning algorithm $\mathfrak{L}$
The number of iterations T

1: $\mathcal{D}_1(\mathbf{x}) = 1/n$ /* initialize the weight distribution */
2: **for** $t = 1$ to T **do**
3:    $h_t = \mathfrak{L}(D, \mathcal{D}_t)$ /* train h_t from D with distribution $\mathcal{D}_t$ */
4:    $\epsilon_t = P_{\mathbf{x} \sim \mathcal{D}_t}(h_t(\mathbf{x}) \neq y(\mathbf{x}))$ /* evaluate the error of h_t */
5:    **if** $\epsilon_t > 0.5$ **then**
6:      **break**
7:    **end if**
8:    $\alpha_t = \frac{1}{2} \ln\left(\frac{1-\epsilon_t}{\epsilon_t}\right)$ /* Determine the weight of h_t */
9:    $\mathcal{D}_{t+1}(\mathbf{x}) = \frac{\mathcal{D}_t(\mathbf{x})}{Z_t} \times \begin{cases} \exp(-\alpha_t) & \text{if } h_t(\mathbf{x}) = y(\mathbf{x}) \\ \exp(\alpha_t) & \text{if } h_t(\mathbf{x}) \neq y(\mathbf{x}) \end{cases}$
          $= \frac{\mathcal{D}_t(\mathbf{x}) \exp(-\alpha_t f(\mathbf{x}) h_t(\mathbf{x}))}{Z_t}$ /* update the distribution */
10: **end for**
Output: $H(\mathbf{x}) = \text{sign}\left(\sum_{t=1}^T \alpha_t h_t(\mathbf{x})\right)$

---

on the mistakes of h_1. Then a base learner h_2 is trained from $\mathcal{D}'$. Again, suppose that h_2 is also a weak learner, and the combination of h_1 and h_2 is not good enough. Boosting will then derive a new distribution $\mathcal{D}''$ to focus on the mistakes of the combined classifier of the former base learners. The algorithm is shown in the following.

It is obvious that the base learners of boosting are dependent, and they are generated in a sequential way. This is different from parallel ensemble methods such as Bagging. Parallel ensemble methods exploit the independency of base learners, while sequential ensemble methods exploit the dependency of base learners.

There are many variants of boosting. The most influential one is AdaBoost [10], which is the best off-the-shelf learning algorithm in the world. AdaBoost can be regarded as an instantiation of the general boosting procedure.

4.2.3 Combination Methods

Averaging, voting, and stacking [11–13] are the most popular combination methods for ensemble. Averaging is commonly used for regression tasks, which averages the numeric outputs of base learners. Voting is the most popular and fundamental combination method for classification tasks. Majority voting is the most common voting method. Each learner votes for one class label, and the ensemble will output the class label with more than half of the votes; if there is no class label that receives more than half of the votes, then no prediction will be given. Unlike majority voting, plurality voting does not need that the output class label must be with more than half of the votes, and it always outputs the class label with the most of the votes. Weighted voting considers that different learners perform differently. It empowers stronger learners to influence more in voting.

Different from averaging and voting that use pre-specified combination rules, stacking trains a learner to combine base learners as a general learning procedure. Base learners are called the *first-level learners*, and the combiner is called the *second-level learner* or *meta-learner*. Briefly speaking, base learners are trained from the original training data, while the second-level learner is trained with different training data, where an example takes the base learners' output as input (attribute values), and the original class label as output.

4.3 ENSEMBLE METHODS FOR CLASS IMBALANCE LEARNING

In CIL, the ground-truth level of imbalance is often unknown, and the ground-truth relative importance of the minority class against the majority class is often unknown either. There are many potential variations; so it is not strange that ensemble methods have been popularly adopted to achieve more effective and robust performance, including Bagging, boosting, Random Forests (RF) [19], stacking, and so on. According to what ensemble method is involved, ensemble methods for CIL can be roughly categorized into Bagging-style methods, boosting-based methods, hybrid ensemble methods, and the others.

Existing CIL methods can be improved by adopting ensemble learning. Under-sampling can solve class imbalance problems effectively and efficiently. However, it is wasteful to ignore potential useful information contained in the majority class examples, as it is often expensive to gather training examples. Great benefits will be obtained if the ignored examples are exploited appropriately. On the other hand, ensemble methods can benefit from the diversity of data. So, it is natural to use ensemble methods to further explore the examples ignored by under-sampling. Several ensemble methods are devoted to improve under-sampling, including Chan and Stolfo's method [5], EasyEnsemble and BalanceCascade [6], UnderBagging [20], and many others.

Over-sampling replicates the majority class examples; so it has the risk of over-fitting. Some ensemble methods are designed to help reduce the risk. For example, DataBoost-IM [8] uses boosting architecture to identify hard examples in each round and creates synthetic examples for the minority class according to the level of imbalance of hard examples. Synthetic minority over-sampling technique (SMOTE) [21] is a state-of-the-art CIL method that reduces the risk of over-fitting by introducing synthetic examples instead of making replicates, but it could introduce noise. To improve SMOTE, SMOTEBoost [4] is proposed. In each boosting round, the weight distribution is adjusted by SMOTE to focus more on the minority class examples. Thus, the weak learners that could be affected by noise can be boosted into strong learners via ensemble.

In the following, we introduce some typical ensemble methods for CIL.

4.3.1 Bagging-Style Methods

Bagging-style CIL methods use diverse training samples (we call them *bags* for convenience) to train independent base learners in parallel, which is like Bagging. But the difference is that, Bagging-style CIL methods try to construct a balanced data sample to train a base learner capable to handle imbalanced data in each iteration, while Bagging uses a bootstrap sample, whose data distribution is identical to the underlying data distribution $\mathcal{D}$, to train a base learner maximizing the accuracy. Different bag construction strategies lead to different Bagging-style methods. Some methods use under-sampling methods to reduce the majority class examples in a bag, such as UnderBagging [20]; some methods use over-sampling methods to increase the minority class examples, such as OverBagging [20] and SMOTEBagging [20]; some methods use the hybrid of under- and/or over-sampling methods to balance the data, such as SMOTEBagging [20]. While some other methods partition the majority class into a set of disjoint subsets of size n_+, with each subset together with all the minority class examples to construct a bag, such as Chan and Stolof's method [5].

4.3.1.1 UnderBagging/OverBagging Both UnderBagging and OverBagging construct a bag by obtaining two samples of the same size $n_- \cdot a\%$ from the minority class and the majority class separately by sampling with replacement, where $a\%$ varies from 10% to 100% [20]. When $a\% = n_+/n_-$, under-sampling is conducted to remove the majority class examples, which leads to UnderBagging; when $a\% = 100\%$, over-sampling is conduced to increase the minority examples, which leads to OverBagging; otherwise, both under- and over-sampling are conduced, which leads to a hybrid Bagging-style ensemble. The base learners are combined by majority voting.

4.3.1.2 SMOTEBagging SMOTEBagging is similar to OverBagging, a sample of n_- minority class examples are sampled [20]. The difference lies in how the sample of minority examples is obtained. SMOTEBagging samples $n_- \cdot b\%$ minority class examples from the minority class and then generates $n_- \cdot (1 - b\%)$ synthetic minority class examples by SMOTE [21].

4.3.1.3 Chan and Stolof's Method Chan and Stolof's method (Chan) partitions the majority class into a set of nonoverlapping subsets, with each subset having approximately n_+ examples [5]. Each of the majority class subset and all the minority class examples form a bag. The base learners are ensembled by stacking.

4.3.1.4 Balanced Random Forests (BRF) [22] Balanced Random Forests (BRF) adapts Random Forests (RF) [19] to imbalanced data. To learn a single tree in each iteration, it first draws a bootstrap sample of size n_+ from the minority class, and then draws the same number of examples with replacement from the majority class. Thus, each training sample is a balanced data sample. Then, BRF induces a tree from each of the balanced training sample with maximum size, without pruning. The tree is generated by classification and regression tree (CART) algorithm [23] with the modification of a random feature subset to split on. The random tree learning procedure and the combination method are the same as in RF. Note that, RF can be used as a normal learning algorithm after balancing training data by under-sampling or over-sampling. But in this procedure, ensemble method is not adopted to handle imbalanced data; so it is different from BRF.

Bagging-style class imbalance methods also have other variants, such as [24, 25]. These methods are very similar to the methods introduced earlier.

4.3.2 Boosting-Based Methods

Boosting-based CIL methods focus more on the minority class examples by increasing the number of minority class examples or decreasing the number of the majority class examples in each boosting round, such as SMOTEBoost [4] and RUSBoost [7], and/or by balancing the weight distribution directly, such as DataBoost-IM [8].

4.3.2.1 SMOTEBoost   SMOTEBoost adopts boosting architecture (AdaBoost. M2) to improve SMOTE, a very popular over-sampling method [4]. In each round of boosting, SMOTE is used to generate synthetic examples for the minority class, and then the weight distribution is adjusted accordingly. Thus, the weights of the minority class examples become higher. It can be used in multiclass cases. The algorithm is shown as follows.

4.3.2.2 RUSBoost   RUSBoost is very similar to SMOTEBoost; the only difference between them is that RUSBoost uses random under-sampling to remove the majority class examples in each boosting round [7].

4.3.2.3 DataBoost-IM   DataBoost-IM designs two strategies to focus not only on the misclassified examples but also on the minority class examples [8]. One is selecting "hard" examples (examples easier to be misclassified) from the majority class and the minority class separately and generating synthetic examples to add into the training data; the other strategy is balancing the total weights of the majority class and the minority class. In detail, in each round, first $N_s = n \times err$ examples with highest weights in current training data are selected, where err is error rate, containing N_{smaj} majority class examples and N_{smin} majority class examples. From these hard examples, $M_L = \min(n_-/n_+, N_{\text{smaj}})$ majority class examples and $M_s = \min((n_- \times M_L)/n_+, N_{\text{smin}})$ minority class examples are selected as seed examples. Then, synthetic examples are generated from the seed examples for the majority class and the minority class separately, each with an initial weight equaling the seed example's weight divided by the number of synthetic examples it generates. Then, the synthetic examples are added into the training data. Thus, the hard minority class examples are more than the hard majority class examples. Finally, the total weights of the majority

---

Algorithm The SMOTEBoost algorithm

Input:

Data set $D = \{(\mathbf{x}_i, y_i)\}_{i=1}^n$ with $y \in \{1, ..., M\}$, where C_m is the minority class

The number of synthetic examples to be generated in each iteration N

The number of iterations T

1: Let $B = \{(i, y) : i = 1, ..., n, y \neq y_i\}$

2: $\mathcal{D}_1(i, y) = 1/|B|$ for $(i, y) \in B$

3: **for** $i = 1$ to T **do**

4:      Modify distribution $\mathcal{D}_t$ by creating N synthetic examples from C_m using the SMOTE algorithm

5:      Train a weak learner using distribution $\mathcal{D}_t$

6:      Compute weak hypothesis $h_t : \mathcal{X} \times \mathcal{Y} \rightarrow [0, 1]$

7:      Compute the pseudo-loss of hypothesis h_t:

$$e_t = \sum_{(i,y) \in B} \mathcal{D}_t(i, y)(1 - h_t(\mathbf{x}_i, y_i) + h_t(\mathbf{x}_i, y))$$

8:      Set $\alpha_t = \ln \frac{1-e_t}{e_t}$

9:      Set $d_t = \frac{1}{2}(1 - h_t(\mathbf{x}_i, y) + h_t(\mathbf{x}_i, y_i))$

10:      Update $\mathcal{D}_t : \mathcal{D}_{t+1}(i, y) = \frac{\mathcal{D}_t(i,y)}{Z_t} e^{-\alpha_t d_t}$

           /* Z_t is a normalization constant such that $\mathcal{D}_{t+1}$ is a distribution */

11: **end for**

Output: $H(\mathbf{x}) = \arg\max_{y \in \mathcal{Y}} \sum_{t=1}^{T} \alpha_t h_t(\mathbf{x}, y)$

---

class and the minority class are balanced, which increases the individual weights of all minority class examples.

4.3.3 Hybrid Ensemble Methods

It is well known that boosting mainly reduces bias, whereas bagging mainly reduces variance. Several methods [26–29] combine different ensemble strategies to achieve stronger generalization. For example, MultiBoosting [27, 28] combines boosting with bagging using boosted ensembles as base learners.

Multiple ensemble strategies can also cooperate to handle imbalanced data, such as EasyEnsemble and BalanceCascade [6]. Both of them try to improve under-sampling by adopting boosting and bagging-style ensemble strategies.

EasyEnsemble [6]. The motivation of EasyEnsemble is to reduce the possibility of ignoring potentially useful information contained in the majority class examples while keeping the high efficiency of under-sampling. First, it generates a set of balanced data samples, with each bag containing all minority class examples and a subset of the majority class examples with size n_+. This step is similar to Bagging-style methods for CIL. Then an AdaBoost ensemble is trained

---

Algorithm   The EasyEnsemble algorithm

Input:
   Data set: $D = \{(\mathbf{x}_i, y_i)\}_{i=1}^{n}$ with minority class $\mathcal{P}$ and majority class $\mathcal{N}$
   The number of iterations: T
   The number of iterations to train an AdaBoost ensemble H_i: s_i
1: **for** $i = 1$ to T **do**
2:    $i \Leftarrow i + 1$
3:    Randomly sample a subset $\mathcal{N}_i$ of n_+ examples form the majority class
4:    Learn H_i using $\mathcal{P}$ and $\mathcal{N}_i$. H_i is an AdaBoost ensemble with s_i weak
   classifiers $h_{i,j}$ and corresponding weights $\alpha_{i,j}$. The ensemble's threshold
   is θ_i, i.e.
$$H_i(x) = \text{sign}\left(\sum_{j=1}^{s_i} \alpha_{i,j} h_{i,j}(\mathbf{x}) - \theta_i\right).$$
5: **end for**
Output: $H(\mathbf{x}) = \text{sign}\left(\sum_{i=1}^{T} \sum_{j=1}^{s_i} \alpha_{i,j} h_{i,j}(\mathbf{x}) - \sum_{i=1}^{T} \theta_i\right)$

---

from each bag. The final ensemble is formed by combining all base learners in all AdaBoost ensembles. The algorithm is shown in the following.

One possible way is to combine all AdaBoost classifiers' predictions, that is,

$$F(\mathbf{x}) = \text{sign}\left(\sum_{i=1}^{T} \text{sign}\left(\sum_{j=1}^{s_i} \alpha_{i,j} h_{i,j}(\mathbf{x}) - \theta_i\right)\right). \tag{4.1}$$

If the ensemble was formed in this way, then the method is just a Bagging-style ensemble method with AdaBoost ensembles as base learners. It is worth noting that it is not the AdaBoost classifiers but the base learners forming them that are combined for prediction, that is,

$$H(\mathbf{x}) = \text{sign}\left(\sum_{i=1}^{T} \sum_{j=1}^{s_i} \alpha_{i,j} h_{i,j}(\mathbf{x}) - \sum_{i=1}^{T} \theta_i\right). \tag{4.2}$$

A different combination method leads to a totally different ensemble method. Some experimental results showed that combining AdaBoost classifiers directly using Equation 4.1 caused an obvious decrease in performance. This is because each AdaBoost classifier gives only a single output, ignoring the detailed information provided by its base learners. The way AdaBoost is used in EasyEnsemble is similar to the usage of AdaBoost in [30], where the base learners are treated as features with binary values. EasyEnsemble also treats the base learners as features to explore different aspects in the learning process.

Algorithm   The BalanceCascade algorithm

Input:
   Data set: $D = \{(\mathbf{x}_i, y_i)\}_{i=1}^n$ with minority class $\mathcal{P}$ and majority class $\mathcal{N}$
   The number of iterations: T
   The number of iterations to train an AdaBoost ensemble H_i: s_i
1: $f \Leftarrow {}^{T-1}\sqrt{\frac{n_+}{n_-}}$, f is the false positive rate that H_i should achieve.
   /* false positive rate is the error rate of misclassifying a majority class example to the minority class */
2: **for** $i = 1$ to T **do**
3:    Randomly sample a subset $\mathcal{N}_i$ of n_+ examples from the majority class
4:    Learn H_i using $\mathcal{P}$ and $\mathcal{N}_i$. H_i is an AdaBoost ensemble with s_i weak classifiers $h_{i,j}$ and corresponding weights $\alpha_{i,j}$. The ensemble's threshold is θ_i i.e.
$$H_i(x) = \mathtt{sign}\left(\sum_{j=1}^{s_i} \alpha_{i,j} h_{i,j}(\mathbf{x}) - \theta_i\right).$$
5:    Adjust θ_i such that H_i's false positive rate is f.
6:    Remove from $\mathcal{N}$ all examples that are correctly classified by H_i.
7: **end for**
8: **return** $H(\mathbf{x}) = \mathtt{sign}\left(\sum_{i=1}^{T} \sum_{j=1}^{s_i} \alpha_{i,j} h_{i,j}(\mathbf{x}) - \sum_{i=1}^{T} \theta_i\right)$

4.3.3.1 BalanceCascade   BalanceCascade tries to delete examples in the majority class in a guided way [6]. Different from EasyEnsemble (which generates subsamples of the majority class in an unsupervised parallel manner), Balance-Cascade works in a supervised sequential manner. The basic idea is to shrink the majority class step by step in cascade style. In each iteration, a subset $\mathcal{N}_i$ of n_+ examples are sampled from the majority class. Then, an AdaBoost ensemble H_i is trained from the union of $\mathcal{N}_i$ and $\mathcal{P}$. After that, the majority class examples that are correctly classified by H_i are considered as redundant information and are removed from the majority class. The final ensemble is formed by combining all the base learners in all the AdaBoost ensembles, as in EasyEnsemble. The algorithm is shown as follows.

4.3.4   Other Ensemble Methods

There are also some ensemble methods that combine different CIL methods to handle imbalanced data. The most feasible way is to ensemble classifiers generated by different methods directly. For example, Zhou and Liu [31] combined the NN classifiers generated by over-sampling, under-sampling, and threshold-moving via hard ensemble and soft ensemble. Some ensemble methods combine classifiers trained from data with different levels of imbalance. For example, Estabrooks et al. [32] generated multiple versions of training data with different levels of imbalance, then trained classifiers from each of them by over- and under-sampling, and ensemble them finally.

In [31], the over-sampling method in use is random over-sampling with replacement. The under-sampling method in use is an informed sampling method, which first removes redundant examples and then removes borderline examples and examples suffering from the class label noise. Redundant examples are the training examples whose role can be replaced by other training examples. They are identified by the 1-NN rule. Borderline examples are the examples close to the boundaries between different classes. They are unreliable because even a small amount of attribute noise can cause the example to be misclassified. The borderline examples and examples suffering from the class label noise are detected by Tomek [33] links. Although threshold-moving is not as popular as sampling methods, it is very important for CIL. It has been stated that trying other methods, such as sampling, without trying by simply setting the threshold may be misleading [34]. The threshold-moving method uses the original training set to train an NN and then moves the decision threshold such that the minority class examples are easier to be predicted correctly. The three methods mentioned earlier are used to train three classifiers that are able to handle imbalanced data, and then hard ensemble and soft ensemble, two popular combination methods, are used to combine them separately. Hard ensemble uses the crisp classification decisions to vote, while soft ensemble uses the normalized real-value outputs to vote.

As shown in previous chapters, cost-sensitive learning methods can be used to handle imbalanced data by assigning higher costs to the minority class examples, so that they can be easily classified correctly. There are many cost-sensitive ensemble methods, especially boosting-based methods. Some methods, such as CBS1, CBS2 [35], and AsymBoost [2], modify the weight-distribution-updating rule, so that the weights of expensive examples are higher. Some methods, such as linear asymmetric classifier (LAC) [30], change the weights of the base learners when forming the ensemble. Some methods, such as AdaC1, AdaC2, AdaC3 [36], and AdaCost [37], not only change the weight-updating rule, but also change the weights of base learners when forming ensemble, by associating the cost with the weighted error rate of each class. Moreover, some methods directly minimize a cost-sensitive loss function, such as Asymmetric Boosting [38].

For example, suppose that the cost of misclassifying a positive and a negative example is $cost_+$ and $cost_-$, respectively. AsymBoost modifies the weight distribution to

$$\mathcal{D}_{t+1}(i) = C\mathcal{D}_t(i)e^{-\alpha_t y_i h_t(\mathbf{x}_i)}$$

$$C = \begin{cases} \sqrt[T]{K}, & \text{for positive examples} \\ 1/\sqrt[T]{K}, & \text{for negative examples} \end{cases}$$

where $K = cost_+/cost_-$ is the cost ratio.

4.4 EMPIRICAL STUDY

Many research works reported the effectiveness of ensemble methods to handle imbalanced data. To illustrate the advantages of ensemble methods for CIL, we compare some typical ones with standard methods and some CIL methods without ensemble techniques.[3]

The datasets are 10 binary class-imbalanced University of California, Irvine (UCI) datasets [39] whose information is summarized in Table 4.1. The methods in comparison are:

1. *Bagging*, with CART as base learning algorithm.
2. *AdaBoost*, with CART as base learning algorithm.
3. *RF* [19]. RF is a state-of-the-art ensemble method. It injects random-ness into base learning algorithm instead of training data. Specifically, RF trains random decision trees as base learners by random feature selection. When constructing a component decision tree, at each step of split selec-tion, RF first selects a feature subset randomly, and then carries out the conventional split selection procedure within the selected feature subset.
4. *Under-sampling + AdaBoost.*   First random under-sampling is used such that the majority class has the same number of examples as the minority class, and then an AdaBoost ensemble is trained from the new dataset.
5. *Under-sampling + RF.*   First random under-sampling is used such that the majority class has the same number of examples as the minority class, and then a RF ensemble is trained from the new dataset.

Table 4.1 Basic Information of Datasets[a]

Dataset	Size	Attribute	Target	#min/#maj	Ratio
abalone	4177	1N,7C	Ring = 7	391/3786	9.7
balance	625	4C	Balance	49/576	11.8
cmc	1473	3B,4N,2C	class 2	333/1140	3.4
aberman	306	1N,2C	class 2	81/225	2.8
housing	506	1B,12C	[20, 23]	106/400	3.8
mf-morph	2000	6C	class 10	200/1800	9.0
mf-zernike	2000	47C	class 10	200/1800	9.0
pima	768	8C	class 1	268/500	1.9
vehicle	846	18C	opel	212/634	3.0
wpbc	198	33C	recur	47/151	3.2

[a] *Size* is the number of examples. *Target* is used as the minority class, and all others are used as the majority class. In *Attribute*, *B*: binary, *N*: nominal, *C*: continuous. #min/#maj is the size of minority and majority classes, respectively. *Ratio* is the size of the majority class divided by that of the minority class.

---

[3] The results are mainly from [6].

6. *Over-sampling + AdaBoost.* First random over-sampling is used such that the minority class has the same number of examples as the majority class, and then an AdaBoost ensemble is trained from the new dataset.

7. *Over-sampling + RF.* First random over-sampling is used such that the minority class has the same number of examples as the majority class, and then a RF ensemble is trained from the new dataset.

8. *SMOTE + under-sampling + AdaBoost.* First SMOTE is used to generate n_+ synthetic examples for the minority class, and then under-sampling is conducted to make the majority class to have $2n_+$ examples; finally, an AdaBoost ensemble is trained from the new dataset.

9. *Chan + AdaBoost.* Chan is a bagging-style ensemble method. First Chan is used to generate independent balanced data samples, and then an AdaBoost ensemble is trained from each sample.

10. *BRF.* This is a bagging-style ensemble method for CIL.

11. *AsymBoost,* with CART as base learning algorithm. AsymBoost is a cost-sensitive boosting method. Let the costs of the positive examples be the level of imbalance, that is, $r = n_-/n_+$, and the costs of the negative examples be 1.

12. *SMOTEBoost,* with CART as base learning algorithm. SMOTEBoost is a boosting method for CIL. The k-nearest neighbor parameter of SMOTE is 5. The amount of new data generated using SMOTE in each iteration is n_+.

13. *EasyEnsemble,* with CART as base learning algorithm for AdaBoost. This is a hybrid ensemble method for CIL.

14. *BalanceCascade,* with CART as base learning algorithm for AdaBoost. This is a hybrid ensemble method for CIL.

For fair comparison, all methods are ensemble methods and they use AdaBoost or CART as base learning algorithm. In addition, the number of all CART base learners trained by these methods (except Chan) is set to 40. As for Chan, there are $\lfloor n_-/n_+ \rfloor$ bags. AdaBoost classifiers are trained for $\lceil 40n_+/n_- \rceil$ iterations when $\lfloor n_-/n_+ \rfloor < 40$; otherwise, only one iteration is allowed. Thus, the number of all CART base learners generated is around 40. The abbreviation and type information for these methods are summarized in Table 4.2, where bold face indicates the method is an ensemble method for CIL. These methods are categorized into three groups: standard ensemble methods (Group 1), CIL methods that do not use ensemble methods to handle imbalanced data (Group 2), and ensemble methods for CIL (Group 3)[4]. Note that although ensemble learning is invoked by the methods in Group 2, it is not a part of CIL methods to handle imbalanced data, which is totally different from the methods in Group 3.

---

[4]We will use "methods in Group 1," "methods in Group 2," and "methods in Group 3" to indicate the above three groups of methods for convenience.

Table 4.2   The Methods in Comparison

		Methods	Abbr.	Type
Group 1	1	Bagging	**Bagg**	Standard ensemble method
	2	AdaBoost	**Ada**	Standard ensemble method
	3	Random Forests	**RF**	Standard ensemble method
Group 2	4	Under-sampling + AdaBoost	**Under-Ada**	CIL method
	5	Under-sampling + Random Forests	**Under-RF**	CIL method
	6	Over-sampling + AdaBoost	**Over-Ada**	CIL method
	7	Over-sampling + Random Forests	**Over-RF**	CIL method
	8	SMOTE + under-sampling + AdaBoost	**SMOTE**	CIL method
Group 3	9	**Chan + AdaBoost**	**Chan**	Bagging-style method for CIL
	10	**Balanced Random Forests**	**BRF**	Bagging-style method for CIL
	11	**AsymBoost**	**Asym**	Boosting-based method for CIL
	12	**SMOTEBoost**	**SMB**	Boosting-based method for CIL
	13	**EasyEnsemble**	**Easy**	Hybrid ensemble method for CIL
	14	**BalanceCascade**	**Cascade**	Hybrid ensemble method for CIL

Table 4.3 summarizes the results of area under the curve (AUC) values [40] by conducting 10 times 10-fold cross-validation. AUC is a popular performance measure in CIL. The higher the AUC value, the better the performance. Figures 4.1 and 4.2 show the scatter plots of methods in Groups 1 and 2 versus each of the six ensemble methods for CIL in Group 3, respectively. A point above the dotted line indicates that the method of y-axes is better than that of the x-axes. In addition, Table 4.4 gives the detailed win–tie–lose counts of each class imbalance method (Groups 2 and 3) versus standard methods (Group 1), and each ensemble method for CIL (Group 3) versus methods in Group 2, via t-tests with significance level at 0.05. Bold face indicates the result is significant in sign test at significance level at 0.05.

Almost all the CIL methods have significantly better performance than CART [6][5]. But when compared with standard ensemble methods (Group 1), under-sampling and over-sampling methods in Group 2 are not very effective. It is probably because standard ensemble methods have strong generalization ability; this could reduce the effect of class imbalance. While **SMOTE** is significantly better than the standard ensemble methods. This suggests that SMOTE sampling and combination of different sampling methods are good choices to handle imbalanced data when invoking ensemble learning algorithm to generate a classifier.

---

[5]Since CART cannot produce AUC values, we did not include CART in comparison list in this chapter. Liu et al. [6] showed that almost all the CIL methods have significantly higher F-measure and G-mean values than CART.

Table 4.3 AUC Results[a]

	abalo.	balan.	cmc	haber.	housi.	mfm	mfz	pima	vehic.	wpbc	avg.
Bagg	0.824	0.439	0.705	0.669	0.825	0.887	0.855	0.821	0.859	0.688	0.757
Ada	0.811	0.616	0.675	0.641	0.815	0.888	0.795	0.788	0.854	0.716	0.760
RF	0.827	0.435	0.669	0.645	0.828	0.880	0.840	0.821	0.869	0.677	0.749
Under-Ada	0.830	0.617	0.671	0.646	0.805	0.916	0.881	0.789	0.846	0.694	0.769
Under-RF	0.842	0.593	0.676	0.643	0.820	0.919	0.889	0.818	0.855	0.661	0.772
Over-Ada	0.817	0.540	0.675	0.637	0.821	0.889	0.779	0.791	0.855	0.711	0.751
Over-RF	0.823	0.458	0.660	0.641	0.826	0.881	0.854	0.819	0.866	0.670	0.750
SMOTE	0.831	0.617	0.680	0.647	0.816	0.912	0.862	0.792	0.858	0.709	0.772
Chan	0.850	0.652	0.696	0.638	0.811	0.912	0.903	0.786	0.856	0.706	0.781
BRF	0.853	0.558	0.683	0.677	0.798	0.901	0.866	0.809	0.850	0.646	0.764
Asym	0.812	0.619	0.675	0.639	0.815	0.888	0.801	0.788	0.853	0.721	0.761
SMB	0.818	0.599	0.687	0.646	0.824	0.897	0.788	0.790	0.864	0.720	0.763
Easy	0.847	0.633	0.704	0.668	0.825	0.918	0.904	0.809	0.859	0.707	0.787
Cascade	0.828	0.637	0.686	0.653	0.808	0.905	0.891	0.799	0.856	0.712	0.778

[a]Data is adapted from [6].

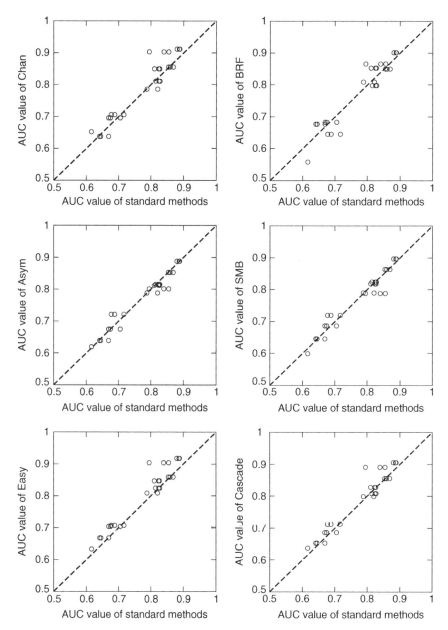

Figure 4.1   Scatter plots of standard methods in Group 1 versus each of the six ensemble methods for class imbalance learning in Group 3.

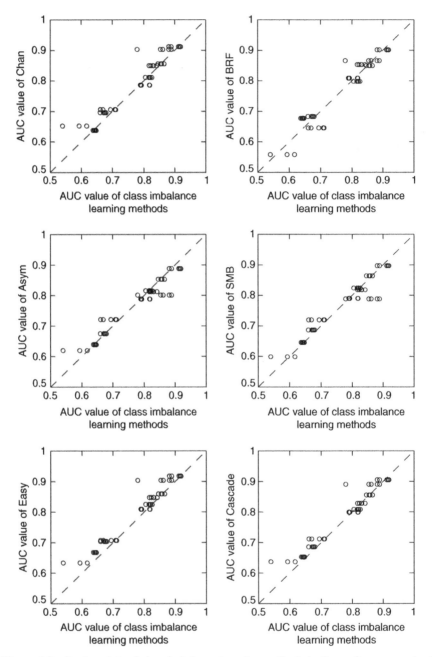

Figure 4.2 Scatter plots of class imbalance learning methods in Group 2 versus each of the six ensemble methods for class imbalance learning in Group 3.

Table 4.4 *t*-Test Results[a,b]

	Methods in Group 1	Methods in Group 2
Under-Ada	12-7-11	—
Under-RF	14-6-10	—
Over-Ada	9-7-14	—
Over-RF	8-11-11	—
SMOTE	15-8-7	—
Chan	15-7-8	**26-12-12**
BRF	17-0-13	24-4-22
Asym	7-11-12	12-15-23
SMB	13-7-10	18-18-14
Easy	**20-7-3**	**40-7-3**
Cascade	17-5-8	**29-9-12**

[a] Data is derived from [6].
[b] The table shows the win–tie–lose counts of method in row versus method in column.

Ensemble methods for CIL (Group 3) are generally better than standard methods (Group 1) and methods in Group 2, except Asym. The bad performance of Asym may be due to the fact that it is a heuristic boosting method designed for cost-sensitive problems. Among these methods, Chan, Easy, and Cascade perform better than the others.

Among these CIL methods, SMB and over-sampling are most time consuming, as the data samples they use to generate classifiers are the largest. While BRF, Chan, Easy, and Cascade are least time consuming, they are as efficient as under-sampling.

It is worth noting that although stacking is a popular combination strategy for ensemble learning, it could be harmful when handling imbalanced data. Liu et al. [6] reported a significant decrease in the performance of EasyEnsemble and BalanceCascade when they use stacking to combine base learners. In imbalanced data, the minority class examples are often rare when they are used for multiple times as in EasyEnsemble and BalanceCascade; stacking has a great chance to over-fit.

4.5 CONCLUDING REMARKS

Ensemble learning is an important paradigm in machine learning, which uses a set of classifiers to make predictions. The generalization ability of an ensemble is generally much stronger than individual ensemble members.

Ensemble learning is very useful when designing CIL methods. The ensemble methods are roughly categorized into Bagging-style methods, such as UnderBagging, OverBagging, SMOTEBagging, and Chan; boosting-based methods, such as SMOTEBoost, RUSBoost, and DataBoost-IM; and hybrid ensemble methods, such as EasyEnsemble and BalanceCascade. It should be noted that a CIL method such as under-sampling invoking ensemble learning algorithm to train a classifier is totally different from ensemble methods designed for CIL because the invoked ensemble method plays no role in handling imbalance data, and it is not a part of a CIL method.

Some empirical results suggested:

1. Compared with standard ensemble methods, under-sampling and over-sampling are not effective in general when invoking ensemble learning algorithm such as AdaBoost to train a classifier. SMOTE sampling and combination of different sampling methods are good choices in such cases.

2. Many of the ensemble methods for CIL are significantly better than standard ensemble methods and sampling-based CIL methods, especially Chan, EasyEnsemble, and BalanceCascade.

3. Stacking may be harmful when used to handle imbalanced data. It has a high risk of over-fitting when the minority class examples are rare and are used for multiple times.

Most of the ensemble methods for CIL deal with binary class problems. How to use ensemble learning to help multiclass problems is an interesting direction. Besides, although the goal of CIL is to achieve higher AUC value, F-measure, or G-mean, most of the methods do not maximize them directly. Using ensemble methods to tackle this problem is a promising direction.

REFERENCES

1. F.-J. Huang, Z.-H. Zhou, H.-J. Zhang, and T. Chen, "Pose invariant face recognition," in *Proceedings of the 4th IEEE International Conference on Automatic Face and Gesture Recognition*, (Grenoble, France), pp. 245–250, 2000.

2. P. Viola and M. Jones, "Fast and robust classification using asymmetric AdaBoost and a detector cascade," in *Advances in Neural Information Processing Systems 14* (T. G. Dietterich, S. Becker, and Z. Ghahramani, eds.), pp. 1311–1318, Cambridge, MA: MIT Press, 2002.

3. P. Viola and M. Jones, "Robust real-time face detection," *International Journal of Computer Vision*, vol. 57, no. 2, pp. 137–154, 2004.

4. N. V. Chawla, A. Lazarevic, L. O. Hall, and K. W. Bowyer, "SMOTEBoost: Improving prediction of the minority class in boosting," in *Proceedings of 7th European Conference on Principles and Practice of Knowledge Discovery in Databases*, (Cavtat-Dubrovnik, Croatia), pp. 107–119, 2003.

5. P. K. Chan and S. J. Stolfo, "Toward scalable learning with non-uniform class and cost distributions: A case study in credit card fraud detection," in *Proceedings of the 4th ACM SIGKDD International Conference on Knowledge Discovery and Data Mining*, (New York), pp. 164–168, 1998.

6. X.-Y. Liu, J. Wu, and Z.-H. Zhou, "Exploratory undersampling for class-imbalance learning," *IEEE Transactions on Systems, Man, and Cybernetics - Part B: Cybernetics*, vol. 39, no. 2, pp. 539–550, 2009.

7. C. Seiffert, T. Khoshgoftaar, J. V. Hulse, and A. Napolitano, "RUSBoost: A hybrid approach to alleviating class imbalance," *IEEE Transactions on Systems, Man, and Cybernetics - Part A: Systems and Humans*, vol. 40, no. 1, pp. 185–197, 2010.

8. H. Guo and H. Viktor, "Learning from imbalanced data sets with boosting and data generation: The Databoost-IM approach," *ACM SIGKDD Explorations Newsletter*, vol. 6, no. 1, pp. 30–39, 2004.

9. L. Breiman, "Bagging predictors," *Machine Learning*, vol. 24, no. 2, pp. 123–140, 1996.

10. Y. Freund and R. E. Schapire, "A decision-theoretic generalization of on-line learning and an application to boosting," *Journal of Computer and System Sciences*, vol. 55, no. 1, pp. 119–139, 1997.

11. D. H. Wolpert, "Stacked generalization," *Neural Networks*, vol. 5, no. 2, pp. 241–260, 1992.

12. L. Breiman, "Stacked regressions," *Machine Learning*, vol. 24, no. 1, pp. 49–64, 1996.

13. P. Smyth and D. Wolper, "Stacked density estimation," in *Advances in Neural Information Processing Systems 10* (M. I. Jordan, M. J. Kearns, and S. A. Solla, eds.), pp. 668–674, Cambridge, MA: MIT Press, 1998.

14. A. Krogh and J. Vedelsby, "Neural network ensembles, cross validation, and active learning," in *Advances in Neural Information Processing Systems 7* (G. Tesauro, D. S. Touretzky, and T. K. Leen, eds.), pp. 231–238, Cambridge, MA: MIT Press, 1995.

15. Z.-H. Zhou, *Ensemble Methods: Foundations and Algorithms*. Boca Raton, FL: Chapman and Hall/CRC Press, 2012.

16. B. Efron and R. Tibshirani, *An Introduction to the Bootstrap*. New York: Chapman & Hall, 1993.

17. M. Kearns and L. G. Valiant, "Cryptographic limitations on learning Boolean formulae and finite automata," in *Proceedings of the 21st Annual ACM Symposium on Theory of Computing* (Seattle, WA), pp. 433–444, 1989.

18. R. E. Schapire, "The strength of weak learnability," *Machine Learning*, vol. 5, no. 2, pp. 197–227, 1990.

19. L. Breiman, "Random forests," *Machine Learning*, vol. 45, no. 1, pp. 5–32, 2001.

20. S. Wang, K. Tang, and X. Yao, "Diversity exploration and negative correlation learning on imbalanced data sets," in *Proceedings of 2009 International Joint Conference on Neural Networks*, (Atlanta, GA), pp. 3259–3266, 2009.

21. N. V. Chawla, K. W. Bowyer, L. O. Hall, and W. P. Kegelmeyer, "SMOTE: Synthetic minority over-sampling technique," *Journal of Artificial Intelligence Research*, vol. 16, pp. 321–357, 2002.

22. C. Chen, A. Liaw, and L. Breiman, "Using random forest to learn imbalanced data," Tech. Rep., University of California, Berkeley, 2004.

23. L. Breiman, J. Friedman, R. Olshen, and C. Stone, *Classification and Regression Trees*. Belmont, CA: Wadsworth International Group, 1984.

24. D. Tao, X. Tang, X. Li, and X. Wu, "Asymmetric bagging and random subspace for support vector machines-based relevance feedback in image retrieval," *IEEE Transactions on Pattern Analysis and Machine Intelligence*, vol. 28, no. 7, pp. 1088–1099, 2006.

25. S. Hido, H. Kashima, and Y. Takahashi, "Roughly balanced bagging for imbalanced data," *Statistical Analysis and Data Mining*, vol. 2, no. 5–6, pp. 412–426, 2009.

26. J. H. Friedman, "Stochastic gradient boosting," *Computational Statistics and Data Analysis*, vol. 38, no. 4, pp. 367–378, 2002.

27. G. I. Webb, "MultiBoosting: A technique for combining boosting and wagging," *Machine Learning*, vol. 40, pp. 159–196, 2000.

28. G. I. Webb and Z. Zheng, "Multistrategy ensemble learning: Reducing error by combining ensemble learning techniques," *IEEE Transactions on Knowledge and Data Engineering*, vol. 16, no. 8, pp. 980–991, 2004.

29. Y. Yu, Z.-H. Zhou, and K. M. Ting., "Cocktail ensemble for regression," in *Proceedings of the 7th IEEE International Conference on Data Mining*, (Omeha, NE), pp. 721–726, 2007.

30. J. Wu, S. C. Brubaker, M. D. Mullin, and J. M. Rehg, "Fast asymmetric learning for cascade face detection," *IEEE Transactions on Pattern Analysis and Machine Intelligence*, vol. 30, pp. 369–382, 2008.

31. Z.-H. Zhou and X.-Y. Liu, "Training cost-sensitive neural networks with methods addressing the class imbalance problem," *IEEE Transactions on Knowledge and Data Engineering*, vol. 18, no. 1, pp. 63–77, 2006.

32. A. Estabrooks, T. Jo, and N. Japkowicz, "A multiple resampling method for learning from imbalanced data sets," *Computational Intelligence*, vol. 20, no. 1, pp. 18–36, 2004.

33. I. Tomek, "Two modifications of CNN," *IEEE Transactions on System, Man, and Cybernetics*, vol. 6, no. 11, pp. 769–772, 1976.

34. F. Provost, "Machine learning from imbalanced data sets 101," in *Proceedings of the AAAI'2000 Workshop on Imbalanced Data Sets*, 2000.

35. K. M. Ting, "A comparative study of cost-sensitive boosting algorithms," in *Proceedings of the 17th International Conference on Machine Learning*, (Stanford, CA), pp. 983–990, 2000.

36. Y. Sun, A. K. C. Wong, and Y. Wang, "Parameter inference of cost-sensitive boosting algorithms," in *Proceedings of the 4th International Conference on Machine Learning and Data Mining in Pattern Recognition*, (Leipzig, Germany), pp. 642–642, 2005.

37. W. Fan, S. J. Stolfo, J. Zhang, and P. K. Chan, "AdaCost: Misclassification cost-sensitive boosting," in *Proceedings of the 16th International Conference on Machine Learning*, (Bled, Slovenia), pp. 97–105, 1999.

38. H. Masnadi-Shirazi and N. Vasconcelos, "Asymmetric boosting," in *Proceedings of the 24th International Conference on Machine Learning*, (Corvalis, Oregon), pp. 609–61, 2007.

39. C. Blake, E. Keogh, and C. J. Merz, "UCI repository of machine learning databases," Department of Information and Computer Science, University of California, Irvine, CA, http://www.ics.uci.edu/~mlearn/MLRepository.html.

40. A. P. Bradley, "The use of the area under the ROC curve in the evaluation of machine learning algorithms," *Pattern Recognition*, vol. 30, no. 6, pp. 1145–1159, 1997.

5

CLASS IMBALANCE LEARNING METHODS FOR SUPPORT VECTOR MACHINES

RUKSHAN BATUWITA
National ICT Australia Ltd., Sydney, Australia

VASILE PALADE
Department of Computer Science, University of Oxford, Oxford, UK

Abstract: Support vector machines (SVMs) is a very popular machine learning technique. Despite of all its theoretical and practical advantages, SVMs could produce suboptimal results with imbalanced datasets. That is, an SVM classifier trained on an imbalanced dataset can produce suboptimal models that are biased toward the majority class and have low performance on the minority class, as most of the other classification paradigms. There have been various data preprocessing and algorithmic techniques proposed in the literature to alleviate this problem for SVMs. This chapter aims to review these techniques.

5.1 INTRODUCTION

Support vector machines (SVMs) [1–7] is a popular machine learning technique, which has been successfully applied to many real-world classification problems from various domains. Owing to its theoretical and practical advantages, such as solid mathematical background, high generalization capability, and ability to find global and nonlinear classification solutions, SVMs have been very popular among the machine learning and data-mining researchers.

Imbalanced Learning: Foundations, Algorithms, and Applications, First Edition.
Edited by Haibo He and Yunqian Ma.
© 2013 The Institute of Electrical and Electronics Engineers, Inc. Published 2013 by John Wiley & Sons, Inc.

Although SVMs often work effectively with balanced datasets, they could produce suboptimal results with imbalanced datasets. More specifically, an SVM classifier trained on an imbalanced dataset often produces models that are biased toward the majority class and have low performance on the minority class. There have been various data preprocessing and algorithmic techniques proposed to overcome this problem for SVMs. This chapter is dedicated to discuss these techniques. In Section 5.2, we present some background on the SVM learning algorithm. In Section 5.3, we discuss why SVMs are sensitive to the imbalance in datasets. Sections 5.4 and 5.5 present the existing techniques proposed in the literature to handle the class imbalance problem for SVMs. Finally, Section 5.6 summarizes this chapter.

5.2   INTRODUCTION TO SUPPORT VECTOR MACHINES

In this section, we briefly review the learning algorithm of SVMs, which has been initially proposed in [1–3]. Let us consider that we have a binary classification problem represented by a dataset $\{(x_1, y_1), (x_2, y_2), \ldots, (x_l, y_l)\}$, where $x_i \in \Re^n$ represents an n-dimensional data point, and $y_i \in \{-1, 1\}$ represents the label of the class of that data point, for $i = 1, \ldots, l$. The goal of the SVM learning algorithm is to find the optimal separating hyperplane that effectively separates these data points into two classes. In order to find a better separation of the classes, the data points are first considered to be transformed into a higher dimensional feature space by a nonlinear mapping function Φ. A possible separating hyperplane residing in this transformed higher dimensional feature space can be represented by,

$$w \cdot \Phi(x) + b = 0 \tag{5.1}$$

where w is the weight vector normal to the hyperplane. If the dataset is completely linearly separable, the separating hyperplane with the maximum margin (for a higher generalization capability) can be found by solving the following maximal margin optimization problem:

$$\min \left(\frac{1}{2} w \cdot w \right)$$

$$\text{s.t.} \quad y_i (w \cdot \Phi(x_i) + b) \geq 1 \tag{5.2}$$

$$i = 1, \ldots, l$$

However, in most real-world problems, the datasets are not completely linearly separable even though they are mapped into a higher dimensional feature space. Therefore, the constrains in the optimization problem mentioned in Equation 5.2 are relaxed by introducing a set of slack variables, $\xi_i \geq 0$. Then, the soft margin optimization problem can be reformulated as follows:

$$\min_{} \left(\frac{1}{2} w \cdot w + C \sum_{i=1}^{l} \xi_i \right)$$

$$\text{s.t.} \quad y_i (w \cdot \Phi(x_i) + b) \geq 1 - \xi_i \tag{5.3}$$

$$\xi_i \geq 0, \quad i = 1, \ldots, l$$

The slack variables $\xi_i > 0$ hold for misclassified examples, and therefore the penalty term $\sum_{i=1}^{l} \xi_i$ can be considered as a measure of the amount of total misclassifications (training errors) of the model. This new objective function given in Equation 5.3 has two goals. One is to maximize the margin and the other one is to minimize the number of misclassifications (the penalty term). The parameter C controls the trade-off between these two goals. This quadratic optimization problem can be easily solved by representing it as a Lagrangian optimization problem, which has the following dual form:

$$\max_{\alpha_i} \left\{ \sum_{i=1}^{l} \alpha_i - \frac{1}{2} \sum_{i=1}^{l} \sum_{j=1}^{l} \alpha_i \alpha_j y_i y_j \Phi(x_i) \cdot \Phi(x_j) \right\} \tag{5.4}$$

$$\text{s.t.} \quad \sum_{i=1}^{l} y_i \alpha_i = 0, \quad 0 \leq \alpha_i \leq C, \quad i = 1, \ldots, l$$

where α_i are Lagrange multipliers, which should satisfy the following Karush–Kuhn–Tucker (KKT) conditions:

$$\alpha_i (y_i (w \cdot \phi(x_i) + b) - 1 + \xi_i) = 0, \quad i = 1, \ldots, l \tag{5.5}$$

$$(C - \alpha_i)\xi_i - 0, \quad i - 1, \ldots, l \tag{5.6}$$

An important property of SVMs is that it is not necessary to know the mapping function $\phi(x)$ explicitly. By applying a kernel function, such that $K(x_i, x_j) = \phi(x_i) \cdot \phi(x_j)$, we would be able to transform the dual optimization problem given in Equation 5.4 into Equation 5.7

$$\max_{\alpha_i} \left\{ \sum_{i=1}^{l} \alpha_i - \frac{1}{2} \sum_{i=1}^{l} \sum_{j=1}^{l} \alpha_i \alpha_j y_i y_j K(x_i, x_j) \right\} \tag{5.7}$$

$$\text{s.t.} \quad \sum_{i=1}^{l} y_i \alpha_i = 0, \quad 0 \leq \alpha_i \leq C, \quad i = 1, \ldots, l$$

By solving Equation 5.7 and finding the optimal values for α_i, w can be recovered as in Equation 5.8

$$w = \sum_{i=1}^{l} \alpha_i y_i \phi(x_i) \qquad (5.8)$$

and b can be determined from the KKT conditions given in Equation 5.5. The data points having nonzero α_i values are called *support vectors*. Finally, the SVM decision function can be given by:

$$f(x) = \texttt{sign}(w \cdot \Phi(x) + b) = \texttt{sign}\left(\sum_{i=1}^{l} \alpha_i y_i K(x_i, x) + b\right) \qquad (5.9)$$

5.3   SVMs AND CLASS IMBALANCE

Although SVMs often produce effective solutions for balanced datasets, they are sensitive to the imbalance in the datasets and produce suboptimal models. Veropoulos et al. [8], Wu and Chang [9], and Akbani et al. [10] have studied this problem closely and proposed several possible reasons as to why SVMs can be sensitive to class imbalance, which are discussed as follows.

5.3.1   Weakness of the Soft Margin Optimization Problem

It has been identified that the separating hyperplane of an SVM model developed with an imbalanced dataset can be skewed toward the minority class [8], and this skewness can degrade the performance of that model with respect to the minority class. This phenomenon can be explained as follows.

Recall the objective function of the SVM soft margin optimization problem, which was given in Equation 5.3 previously.

$$\min\left(\frac{1}{2}w \cdot w + C\sum_{i=1}^{l} \xi_i\right)$$

$$\text{s.t.} \quad y_i(w \cdot \Phi(x_i) + b) \geq 1 - \xi_i \qquad (5.10)$$

$$\xi_i \geq 0, \quad i = 1, \ldots, l$$

The first part of this objective function focuses on maximizing the margin, while the second part attempts to minimize the penalty term associated with the misclassifications, where the regularization parameter C can also be considered as the assigned misclassification cost. Since we consider the same misclassification cost for all the training examples (i.e., same value of C for both positive and negative examples), in order to reduce the penalty term, the total number of misclassifications should be reduced. When the dataset is imbalanced, the density of majority class examples would be higher than the density of minority class examples even around the class boundary region, where the ideal hyperplane would pass through (throughout this chapter we consider the majority class as the negative class and the minority class as the positive class). It is also pointed out in [9] that the low presence of positive examples makes them appear further from the ideal class boundary than the negative examples. As a consequence, in order to reduce the total number of misclassifications in SVM learning, the separating hyperplane can be shifted (or skewed) toward the minority class. This shift/skew can cause the generation of more false negative predictions, which lowers the model's performance on the minority positive class. When the class imbalance is extreme, the SVMs could produce models having largely skewed hyperplanes, which would even recognize all the examples as negatives [10].

5.3.2   The Imbalanced Support-Vector Ratio

Wu and Chang [9] have experimentally identified that as the training data gets more imbalanced, the ratio between the positive and negative support vectors also becomes more imbalanced. They have hypothesized that as a result of this imbalance, the neighborhood of a test instance close to the boundary is more likely to be dominated by negative support vectors, and hence the decision function is more likely to classify a boundary point as negative. However, Akbani et al. [10] have argued against this idea by pointing out that because of the constraint $\sum_{i=1}^{l} y_i \alpha_i = 0$ (given in Eq. 5.4), α_i of each positive support vector, which are less in numbers than the negative support vectors, must be larger in magnitude than the α_i values associated with the negative support vectors. These α_i act as weights in the final decision function (Eq. 5.9), and hence larger α_i in the positive support vectors receive higher weights than the negative support vectors, which can reduce the effect of imbalance in support vectors up to some extent. Akbani et al. [10] have further argued that this could be the reason why SVMs do not perform very badly, as compared to other machine learning algorithms for moderately skewed datasets.

   In the remaining sections of this chapter, we review the methods found in the literature to handle the class imbalance problem for SVMs. These methods have been developed as both data preprocessing methods (called *external methods*) and algorithmic modifications to the SVM algorithm (called *internal methods*).

5.4   EXTERNAL IMBALANCE LEARNING METHODS FOR SVMs: DATA PREPROCESSING METHODS

5.4.1   Resampling Methods

All the data preprocessing methods discussed in the other chapters of this book can be used to balance the datasets before training the SVM models. These methods include random and focused under/oversampling methods and synthetic data generation methods such as the synthetic minority oversampling technique (SMOTE) [11]. Resampling methods have been successfully applied to train SVMs with imbalanced datasets in different domains [10–16].

Especially, Batuwita and Palade [17] present an efficient focused oversampling method for SVMs. In this method, first the separating hyperplane found by training an SVM model on the original imbalanced dataset is used to select the most informative examples for a given classification problem, which are the data points lying around the class boundary region. Then, only these selected examples are balanced by oversampling as opposed to blindly oversampling the complete dataset. This method reduces the SVM training time significantly while obtaining comparable classification results to the original oversampling method.

Support cluster machines (SCMs) method presented in [18] can be viewed as another focused resampling method for SVMs. This method first partitions the negative examples into disjoint clusters using the kernel-k-means clustering method. Then, it trains an initial SVM model using the positive examples and the representatives of the negative clusters, namely the data examples representing the cluster centers. With the global picture of the initial SVMs, it approximately identifies the support vectors and nonsupport vectors. Then, a shrinking technique is used to remove the samples that are most probably not support vectors. This procedure of clustering and shrinking is performed iteratively several times until convergence.

5.4.2   Ensemble Learning Methods

Ensemble learning has also been applied as a solution for training SVMs with imbalanced datasets [19–22]. Generally, in these methods, the majority class dataset is divided into multiple subdatasets such that each of these sub-datasets has a similar number of examples as the minority class dataset. This can be done by random sampling with or without replacement (bootstrapping), or through clustering methods. Then, a set of SVM classifiers is developed, so that each one is trained with the same positive dataset and a different negative sub-dataset. Finally, the decisions made by the classifier ensemble are combined by using a method such as majority voting. In addition, special boosting algorithms, such as Adacost [23], RareBoost [24], and SMOTEBoost [25], which have been used in class imbalance learning with ensemble settings, could also be applied with SVMs.

5.5   INTERNAL IMBALANCE LEARNING METHODS FOR SVMs: ALGORITHMIC METHODS

In this section, we present the algorithmic modifications proposed in the literature to make the SVM algorithm less sensitive to class imbalance.

5.5.1 Different Error Costs (DEC)

As pointed out in Section 5.3, the main reason for the SVM algorithm to be sensitive to class imbalance would be that the soft margin objective function given in Equation 5.10 assigns the same cost (i.e., C) for both positive and negative misclassifications in the penalty term. This would cause the separating hyperplane to be skewed toward the minority class, which would finally yield a suboptimal model. The different error costs (DEC) method is a cost-sensitive learning solution proposed in [8] to overcome this problem in SVMs. In this method, the SVM soft margin objective function is modified to assign two misclassification costs, such that C^+ is the misclassification cost for positive class examples, while C^- is the misclassification cost for negative class examples, as given in Equation 5.11. Here we also assume positive class to be the minority class and negative class to be the majority class.

$$
\min \left(\frac{1}{2} w \cdot w + C^+ \sum_{i|y_i=+1}^{l} \xi_i + C^- \sum_{i|y_i=-1}^{l} \xi_i \right)
$$
$$
\text{s.t.} \quad y_i(w \cdot \Phi(x_i) + b) \geq 1 - \xi_i \qquad (5.11)
$$
$$
\xi_i \geq 0, \quad i = 1, \ldots, l
$$

By assigning a higher misclassification cost for the minority class examples than the majority class examples (i.e., $C^+ > C^-$), the effect of class imbalance could be reduced. That is, the modified SVM algorithm would not tend to skew the separating hyperplane toward the minority class examples to reduce the total misclassifications, as the minority class examples are now assigned with a higher misclassification cost. The dual Lagrangian form of this modified objective function can be represented as follows:

$$
\max_{\alpha_i} \left\{ \sum_{i=1}^{l} \alpha_i - \frac{1}{2} \sum_{i=1}^{l} \sum_{j=1}^{l} \alpha_i \alpha_j y_i y_j K(x_i, x_j) \right\}
$$
$$
\text{s.t.} \quad \sum_{i=1}^{l} y_i \alpha_i = 0, \quad 0 \leq \alpha_i^+ \leq C^+, \quad 0 \leq \alpha_i^- \leq C^-, \quad i = 1, \ldots, l \quad (5.12)
$$

where α_i^+ and α_i^- represent the Lagrangian multipliers of positive and negative examples, respectively. This dual optimization problem can be solved in the same way as solving the normal SVM optimization problem. As a rule of thumb, Akbani et al. [10] have reported that reasonably good classification results could be obtained from the DEC method by setting the C^-/C^+ equal to the minority-to-majority-class ratio.

5.5.2   One-Class Learning

Raskutti and Kowalczyk [26] and Kowalczyk and Raskutti [27] have presented two extreme rebalancing methods for training SVMs with highly imbalanced datasets. In the first method, they have trained an SVM model only with the minority class examples. In the second method, the DEC method has been extended to assign a $C^- = 0$ misclassification cost for the majority class examples and $C^+ = 1/N^+$ misclassification cost for minority class examples, where N^+ is the number of minority class examples. From the experimental results obtained on several heavily imbalanced synthetic and real-world datasets, these methods have been observed to be more effective than general data rebalancing methods.

5.5.3   zSVM

zSVM is another algorithmic modification proposed for SVMs in [28] to learn from imbalanced datasets. In this method, first an SVM model is developed by using the original imbalanced training dataset. Then, the decision boundary of the resulted model is modified to remove its bias toward the majority (negative) class. Consider the standard SVM decision function given in Equation 5.9, which can be rewritten as follows:

$$f(x) = \text{sign}\left(\sum_{i=1}^{l}\alpha_i y_i K(x_i, x) + b\right)$$

$$= \text{sign}\left(\sum_{i=1}^{l_1}\alpha_i^+ y_i K(x_i, x) + \sum_{j=1}^{l_2}\alpha_j^- y_j K(x_j, x) + b\right) \tag{5.13}$$

where α_i^+ are the coefficients of the positive support vectors, α_j^- are the coefficients of the negative support vectors, and l_1 and l_2 represent the number of positive and negative training examples, respectively. In the zSVM method, the magnitude of the α_i^+ values of the positive support vectors is increased by multiplying all of them by a particular small positive value z. Then, the modified SVM decision function can be represented as follows:

$$f(x) = \text{sign}\left(z * \sum_{i=1}^{l_1}\alpha_i^+ y_i K(x_i, x) + \sum_{j=1}^{l_2}\alpha_i^- y_i K(x_j, x) + b\right) \tag{5.14}$$

This modification of α_i^+ would increase the weights of the positive support vectors in the decision function, and therefore it would decrease its bias toward the majority negative class. In [28], the value of z giving the best classification results for the training dataset was selected as the optimal value.

5.5.4 Kernel Modification Methods

Several techniques have been proposed in the literature to make the SVM algorithm less sensitive to the class imbalance by modifying the associated kernel function.

5.5.4.1 Class Boundary Alignment Wu and Chang [9] have proposed a variant of SVM learning method, where the kernel function is conformally transformed to enlarge the margin around the class boundary region in the transformed higher dimensional feature space in order to have improved performance. Wu and Chang [29] have improved this method for imbalanced datasets by enlarging more of the class boundary around the minority class compared to the class boundary around the majority class. This method is called the *class boundary alignment* (CBA) method, which can only be used with the vector space representation of input data. Wu and Chang [30] have further proposed a variant of the CBA method for the sequence representation of imbalanced input data by modifying the kernel matrix to have a similar effect, which is called the *kernel boundary alignment* (KBA) method.

5.5.4.2 Kernel Target Alignment In the context of SVM learning, a quantitative measure of agreement between the kernel function used and the learning task is important from both the theoretical and practical points of view. Kernel target alignment method has been proposed as a method for measuring the agreement between a kernel being used and the classification task in [31]. This method has been improved for imbalanced datasets learning in [32].

5.5.4.3 Margin Calibration The DEC method described previously modifies the SVM objective function by assigning a higher misclassification cost to the positive examples than the negative examples to change the penalty term. Yang et al. [33] have extended this method to modify the SVM objective function not only in terms of the penalty term, but also in terms of the margin to recover the biased decision boundary. As proposed in this method, the modification first adopts an inversed proportional regularized penalty to reweight the imbalanced classes. Then it employs a margin compensation to lead the margin to be lopsided, which enables the decision boundary drift.

5.5.4.4 Other Kernel-Modification Methods There have been several imbalance learning techniques proposed in the literature for other kernel-based classifiers. These methods include the kernel classifier construction algorithm proposed in [34], based on orthogonal forward selection (OFS) and regularized orthogonal weighted least squares (ROWLSs) estimator, kernel neural gas (KNG) algorithm for imbalanced clustering [35], the P2PKNNC algorithm based on the k-nearest neighbors classifier and the P2P communication paradigm [36], Adaboost relevance vector machine (RVM) [37], among others.

5.5.5 Active Learning

Active learning methods, as opposed to conventional batch learning, have also been applied to solve the problem of class imbalance for SVMs. Ertekin et al. [38, 39] have proposed an efficient active learning strategy for SVMs to overcome the class imbalance problem. This method iteratively selects the closest instance to the separating hyperplane from the unseen training data and adds it to the training set to retrain the classifier. With an early stopping criterion, the method can significantly decrease the training time on large-scale imbalanced datasets.

5.5.6 Fuzzy SVMs for Class Imbalance Learning (FSVM-CIL)

All the methods presented so far attempt to make SVMs robust to the problem of class imbalance. It has been well studied in the literature that SVMs are also sensitive to the noise and outliers present in datasets. Therefore, it can be argued that although the existing class imbalance learning methods can make the SVM algorithm less sensitive to the class imbalance problem, it can still be sensitive to noise and outliers present in datasets, which could still result in suboptimal models. In fact, some class imbalance learning methods, such as random oversampling and SMOTE, can make the problem worse by duplicating the existing outliers and noisy examples or introducing new ones. Fuzzy SVMs for class imbalance learning (FSVM-CIL) is an improved SVM method proposed in [40] to handle the problem of class imbalance together with the problem of outliers and noise. In this section, we present this method in more detail.

5.5.6.1 The Fuzzy SVM Method   As mentioned previously, the standard SVM algorithm considers all the data points with equal importance and assigns the same misclassification cost for those in its objective function. We have already pointed out that this can cause SVM to produce suboptimal models on imbalanced datasets. It has also been found out that the same reason for considering all the data points with equal importance can also cause SVMs to be sensitive to the outliers and noise present in a dataset. That is, the presence of outliers and noisy examples (especially, around the class boundary region) can influence the position and orientation of the separating hyperplane causing the development of suboptimal models.

In order to make the SVMs less sensitive to outliers and noisy examples, a technique called *Fuzzy SVMs (FSVMs)* has been proposed in [41]. The FSVM method assigns different fuzzy membership values, m_i; $m_i \geq 0$ (or weights), for different examples to reflect the different importance in their own classes, where more important examples are assigned higher membership values, while less important ones (such as outliers and noise) are assigned lower membership values. Then, the SVM soft margin optimization problem is reformulated as follows:

$$\min\left(\frac{1}{2}w \cdot w + C\sum_{i=1}^{l} m_i \xi_i\right)$$

$$\text{s.t.} \quad y_i (w \cdot \Phi(x_i) + b) \geq 1 - \xi_i \tag{5.15}$$

$$\xi_i \geq 0, \quad i = 1, \ldots, l$$

In this reformulation of the objective function, the membership m_i of a data point x_i is incorporated into the penalty term, such that a smaller m_i could reduce the effect of the associated slack variable ξ_i in the objective function (if the corresponding data point x_i is treated as less important). In another view, if we consider C as the cost assigned for a misclassification, now each data point is assigned with a different misclassification cost, $m_i C$, which is based on the importance of the data point in its own class, such that more important data points are assigned higher costs, while less important ones are assigned lower costs. Therefore, the FSVM algorithm can find a more robust separating hyperplane through maximizing the margin by allowing some misclassification for less important examples, such as the outliers and noise.

In order to solve the FSVM optimization problem, Equation 5.15 can be transformed into the following dual Lagrangian form:

$$\max_{\alpha_i} \left\{ \sum_{i=1}^{l} \alpha_i - \frac{1}{2} \sum_{i=1}^{l} \sum_{j=1}^{l} \alpha_i \alpha_j y_i y_j K(x_i, x_j) \right\} \tag{5.16}$$

$$\text{s.t.} \quad \sum_{i=1}^{l} y_i \alpha_i = 0, \quad 0 \leq \alpha_i \leq m_i C, \quad i = 1, \ldots, l$$

The only difference between the original SVM dual optimization problem given in Equation 5.7 and the FSVM dual optimization problem given in Equation 5.16 is the upper bound of the values that α_i could take. By solving this dual problem in Equation 5.16 for optimal α_i, w and b can be recovered in the same way as in the normal SVM learning algorithm. The same SVM decision function in Equation 5.9 applies for the FSVMs method as well.

5.5.6.2 FSVM-CIL Method

However, the standard FSVM method is still sensitive to the class imbalance problem, as the assigned misclassification costs do not consider the imbalance of the dataset. Batuwita and Palade [40] have improved the standard FSVM method by combining it with the DEC method, which is called the *FSVM-CIL*. In the FSVM-CIL method, the membership values for data points are assigned in such a way to satisfy the following two goals:

1. To suppress the effect of between-class imbalance.
2. To reflect the within-class importance of different training examples in order to suppress the effect of outliers and noise.

Let m_i^+ represents the membership value of a positive data point x_i^+, while m_i^- represents the membership of a negative data point x_i^- in their own classes. In the proposed FSVM-CIL method, these membership functions are defined as follows:

$$m_i^+ = f(x_i^+)r^+ \tag{5.17}$$

$$m_i^- = f(x_i^-)r^- \tag{5.18}$$

where $f(x_i)$ generates a value between 0 and 1, which reflects the importance of x_i in its own class. The values for r^+ and r^- were assigned in order to reflect the class imbalance, such that $r^+ = 1$ and $r^- = r$, where r is the minority-to-majority-class ratio ($r^+ > r^-$) (this was following the findings reported in [10], where optimal results could be obtained from the DEC method when C^-/C^+ equals to the minority-to-majority-class ratio). According to this assignment of values, a positive class data point is assigned a misclassification cost m_i^+C, where m_i^+ takes a value in the $[0,1]$ interval, while a negative class data point is assigned a misclassification cost m_i^-C, where m_i^- takes value in the $[0, r]$ interval, where $r < 1$.

In order to define the function $f(x_i)$ introduced in Equations 5.17 and 5.18, which gives the within-class importance of a training example, the following methods have been considered in [40].

A. $f(x_i)$ is based on the distance from the own class center: In this method, $f(x_i)$ is defined with respect to d_i^{cen}, which is the distance between x_i and its own class center. The examples closer to the class center are treated as more informative and assigned higher $f(x_i)$ values, while the examples far away from the center are treated as outliers or noise and assigned lower $f(x_i)$ values. Here, two separate decaying functions of d_i^{cen} have been used to define $f(x_i)$, which are represented by $f_{lin}^{cen}(x_i)$ and $f_{exp}^{cen}(x_i)$ as follows:

$$f_{lin}^{cen}(x_i) = 1 - (d_i^{cen}/(\max(d_i^{cen}) + \delta)) \tag{5.19}$$

is a linearly decaying function. δ is a small positive value used to avoid the case where $f(x_i)$ becomes zero.

$$f_{exp}^{cen}(x_i) = 2/(1 + \exp(d_i^{cen} * \beta)) \tag{5.20}$$

is an exponentially decaying function, where β; $\beta \in [0, 1]$ determines the steepness of the decay. $d_i^{cen} = \|x_i - \overline{x}\|^{\frac{1}{2}}$ is the Euclidean distance to x_i from its own class center $\overline{x}$.

B. $f(x_i)$ *is based on the distance from the preestimated separating hyperplane:* In this method, $f(x_i)$ is defined based on d_i^{sph}, which is the distance to x_i from the pre-estimated separating hyperplane, as introduced in [42]. Here d_i^{sph} is estimated by the distance to x_i from the center of the common spherical region, which can be defined as a hyper-sphere covering the overlapping region of the two classes, where the separation hyperplane is more likely to pass through. Both linear and exponential decaying functions are used to define the function $f(x_i)$, which are represented by $f_{lin}^{sph}(x_i)$ and $f_{exp}^{sph}(x_i)$ as follows:

$$f_{lin}^{sph}(x_i) = 1 - (d_i^{sph}/(\max(d_i^{sph}) + \delta)) \qquad (5.21)$$

$$f_{exp}^{sph}(x_i) = 2/(1 + \exp(d_i^{sph} * \beta)) \qquad (5.22)$$

where $d_i^{sph} = \|x_i - \overline{x}\|^{\frac{1}{2}}$, and $\overline{x}$ is the center of the spherical region, which is estimated by the center of the entire dataset, and δ is a small positive value and $\beta \in [0, 1]$.

C. $f(x_i)$ *is based on the distance from the actual separating hyperplane:* In this method, $f(x_i)$ is defined based on the distance from the actual separating hyperplane to x_i, which is found by training a conventional SVM model on the imbalanced dataset. The data points closer to the actual separating hyperplane are treated as more informative and are assigned higher membership values, while the data points far away from the separating hyperplane are treated as less informative and are assigned lower membership values. The following procedure is carried out to assign $f(x_i)$ values in this method:

1. Train a normal SVM model with the original imbalanced dataset
2. Find the functional margin d_i^{hyp} of each example x_i (given in Eq. 5.23) (this is equivalent to the absolute value of the SVM decision value) with respect to the separating hyperplane found. The functional margin is proportional to the geometric margin of a training example with respect to the separating hyperplane.

$$d_i^{hyp} = y_i(w \cdot \Phi(x_i) + b) \qquad (5.23)$$

3. Consider both linear and exponential decaying functions to define $f(x_i)$ as follows:

$$f_{lin}^{hyp}(x_i) = 1 - (d_i^{hyp}/(\max(d_i^{hyp}) + \delta)) \qquad (5.24)$$

$$f_{exp}^{hyp}(x_i) = 2/(1 + \exp(d_i^{hyp} * \beta)) \qquad (5.25)$$

where δ is a small positive value and $\beta \in [0, 1]$.

Following the aforementioned methods of assigning membership values for positive and negative training data points, several FSVM-CIL settings have been defined in [40]. These methods have been validated on 10 real-world imbalanced datasets representing a variety of domains, complexities, and imbalanced ratios, which are highly likely to contain noisy examples and outliers. FSVM-CIL settings have resulted in better classification results on these datasets than the existing class imbalance learning methods applied for standard SVMs, namely random oversampling, random undersampling, SMOTE, DEC, and zSVM methods. Batuwita and Palade [40] pointed out that better performance of FSVM-CIL method is due to its capability to handle outliers and noise in these datasets in addition to the class imbalance problem.

5.5.7 Hybrid Methods

There exist methods that have used the combination of both external and internal methods to solve the class imbalance problem for SVMs. The hybrid kernel machine ensemble (HKME) method [43] combines a standard binary SVM and a one-class SVM classifier to solve the problem of class imbalance. Akbani et al. [10] has combined the SMOTE algorithm with the DEC method for SVMs for imbalanced dataset learning and shown to have better performance than the use of either of these methods alone.

5.6 SUMMARY

This chapter aimed to review the existing imbalance learning methods developed for SVMs. These methods have been developed as data pre-processing methods or algorithmic improvements. As pointed out in the literature, the class imbalance learning method giving the optimal solution is often dataset dependent. Therefore, it is worth applying several of these available external and internal methods and compare the performances, when training an SVM model on an imbalanced dataset.

REFERENCES

1. V. Vapnik, *The Nature of Statistical Learning Theory*. New York: Springer-Verlag, Inc., 1995.
2. C. Cortes and V. Vapnik, "Support-vector networks," *Machine Learning*, vol. 20, no. 3, pp. 273–297, 1995.
3. B. Boser, I. Guyon, and V. Vapnik, "A training algorithm for optimal margin classifiers," in *Proceedings of the 5th Annual ACM Workshop on Computational Learning Theory (Pittsburgh, PA, USA)*, pp. 144–152, ACM Press, 1992.
4. N. Cristianinio and J. Shawe-Taylor, *An Introduction to Support Vector Machines and Other Kernel-Based Learning Methods*. Cambridge: Cambridge University Press, 2000.

5. B. Scholkopf and A. Smola, *Learning with Kernels: Support Vector Machines, Regularization, Optimization, and Beyond*. Cambridge, MA: MIT Press, 2001.

6. C. Burges, "A tutorial on support vector machines for pattern recognition," *Data Mining and Knowledge Discovery*, vol. 2, no. 2, pp. 121–167, 1998.

7. C.-C. Chang and C.-J. Lin, "LIBSVM: A library for support vector machines," *ACM Transactions on Intelligent Systems and Technology*, vol. 2, pp. 1–27, 2011.

8. K. Veropoulos, C. Campbell, and N. Cristianini, "Controlling the sensitivity of support vector machines," in *Proceedings of the International Joint Conference on Artificial Intelligence* (Stockholm, Sweden), pp. 55–60, 1999.

9. G. Wu and E. Chang, "Adaptive feature-space conformal transformation for imbalanced-data learning," in *Proceedings of the 20th International Conference on Machine Learning* (Washington, DC), pp. 816–823, IEEE Press, 2003.

10. R. Akbani, S. Kwek, and N. Japkowicz, "Applying support vector machines to imbalanced datasets," in *Proceedings of the 15th European Conference on Machine Learning* (Pisa, Italy), pp. 39–50, Springer, 2004.

11. N. Chawla, K. Bowyer, L. Hall, and W. Kegelmeyer, "SMOTE: Synthetic minority over-sampling technique," *Journal of Artificial Intelligence Research*, vol. 16, pp. 321–357, 2002.

12. J. Chen, M. Casique, and M. Karakoy, "Classification of lung data by sampling and support vector machine," in *Proceedings of the 26th Annual International Conference of the IEEE Engineering in Medicine and Biology Society* (San Francisco, CA), vol. 2, pp. 3194–3197, 2004.

13. Y. Fu, S. Ruixiang, Q. Yang, H. Simin, C. Wang, H. Wang, S. Shan, J. Liu, and W. Gao, "A block-based support vector machine approach to the protein homology prediction task in kdd cup 2004," *SIGKDD Exploration Newsletters*, vol. 6, pp. 120–124, 2004.

14. S. Lessmann, "Solving imbalanced classification problems with support vector machines," in *Proceedings of the International Conference on Artificial Intelligence* (Las Vegas, NV, USA), pp. 214–220, 2004.

15. R. Batuwita and V. Palade, "An improved non-comparative classification method for human microrna gene prediction," in *Proceedings of the International Conference on Bioinformatics and Bioengineering* (Athens, Greece), pp. 1–6, IEEE Press, 2008.

16. R. Batuwita and V. Palade, "microPred: Effective classification of pre-miRNAs for human miRNA gene prediction," *Bioinformatics*, vol. 25, pp. 989–995, 2009.

17. R. Batuwita and V. Palade, "Efficient resampling methods for training support vector machines with imbalanced datasets," in *Proceedings of the International Joint Conference on Neural Networks* (Barcelona, Spain), pp. 1–8, IEEE Press, 2010.

18. J. Yuan, J. Li, and B. Zhang, "Learning concepts from large scale imbalanced data sets using support cluster machines," in *Proceedings of the 14th Annual ACM International Conference on Multimedia (Santa Barbara, CA, USA)*, pp. 441–450, ACM, 2006.

19. Z. Lin, Z. Hao, X. Yang, and X. Liu, "Several SVM ensemble methods integrated with under-sampling for imbalanced data learning," in *Proceedings of the 5th International Conference on Advanced Data Mining and Applications* (Beijing, China), pp. 536–544, Springer-Verlag, 2009.

20. P. Kang and S. Cho, "EUS SVMS: Ensemble of under-sampled SVMS for data imbalance problems," in *Proceedings of the 13th International Conference on Neural Information Processing* (Hong Kong, China), pp. 837–846, Springer-Verlag, 2006.

21. Y. Liu, A. An, and X. Huang, "Boosting prediction accuracy on imbalanced datasets with svm ensembles," in *Proceedings of the 10th Pacific-Asia conference on Advances in Knowledge Discovery and Data Mining* (Singapore), pp. 107–118, Springer Verlag, 2006.

22. B. Wang and N. Japkowicz, "Boosting support vector machines for imbalanced data sets," *Knowledge and Information Systems*, vol. 25, pp. 1–20, 2010.

23. W. Fan, S. Stolfo, J. Zhang, and P. Chan, "Adacost: Misclassification cost-sensitive boosting," in *Proceedings of the 16th International Conference on Machine Learning* (Bled, Slovenia), pp. 97–105, Morgan Kaufmann Publishers, Inc., 1999.

24. M. Joshi, V. Kumar, and C. Agarwal, "Evaluating boosting algorithms to classify rare classes: Comparison and improvements," in *Proceedings of the IEEE International Conference on Data Mining*, pp. 257–264, IEEE Computer Society, 2001.

25. N. Chawla, A. Lazarevic, L. Hall, and K. Bowyer, "Smoteboost: Improving prediction of the minority class in boosting," in *Proceedings of the Principles of Knowledge Discovery in Databases* (Cavtat-Dubrovnik, Croatia), pp. 107–119, Springer Verlag, 2003.

26. B. Raskutti and A. Kowalczyk, "Extreme re-balancing for SVMS: A case study," *SIGKDD Exploration Newsletters*, vol. 6, pp. 60–69, June 2004.

27. A. Kowalczyk and B. Raskutti, "One class SVM for yeast regulation prediction," *SIGKDD Exploration Newsletters*, vol. 4, no. 2, pp. 99–100, 2002.

28. T. Imam, K. Ting, and J. Kamruzzaman, "z-SVM: An SVM for improved classification of imbalanced data," in *Proceedings of the 19th Australian Joint Conference on Artificial Intelligence: Advances in Artificial Intelligence* (Hobart, Australia), pp. 264–273, Springer-Verlag, 2006.

29. G. Wu and E. Chang, "Class-boundary alignment for imbalanced dataset learning," in *Proceeding of the International Conference on Machine Learning: Workshop on Learning from Imbalanced Data Sets*, pp. 49–56, 2003.

30. G. Wu and E. Chang, "KBA: Kernel boundary alignment considering imbalanced data distribution," *IEEE Transactions on Knowledge and Data Engineering*, vol. 17, no. 6, pp. 786–795, 2005.

31. N. Cristianini, J. Kandola, A. Elisseeff, and J. Shawe-Taylor, "On kernel-target alignment," in *Advances in Neural Information Processing Systems 14*, pp. 367–373, MIT Press, 2002.

32. J. Kandola and J. Shawe-taylor, "Refining kernels for regression and uneven classification problems," in *Proceedings of International Conference on Artificial Intelligence and Statistics*, Springer-Verlag, 2003.

33. C.-Y. Yang, J.-S. Yang, and J.-J. Wang, "Margin calibration in SVM class-imbalanced learning," *Neurocomputing*, vol. 73, no. 1–3, pp. 397–411, 2009.

34. X. Hong, S. Chen, and C. Harris, "A kernel-based two-class classifier for imbalanced data sets," *IEEE Transactions on Neural Networks*, vol. 18, no. 1, pp. 28–41, 2007.

35. A. Qin and P. Suganthan, "Kernel neural gas algorithms with application to cluster analysis," in *Proceedings of the 17th International Conference on Pattern Recognition* (Cambridge, UK), pp. 617–620, IEEE Computer Society, 2004.

36. X.-P. Yu and X.-G. Yu, "Novel text classification based on k-nearest neighbor," in *Proceedings of the International Conference on Machine Learning and Cybernetics* (Hong Kong, China), pp. 3425–3430, 2007.

37. A. Tashk and K. Faez, "Boosted Bayesian kernel classifier method for face detection," in *Proceedings of the Third International Conference on Natural Computation* (Haikou, Hainan, China), pp. 533–537, IEEE Computer Society, 2007.

38. S. Ertekin, J. Huang, and L. Giles, "Active learning for class imbalance problem," in *Proceedings of the 30th Annual International ACM SIGIR Conference on Research and Development in Information Retrieval* (Amsterdam, The Netherlands), pp. 823–824, ACM, 2007.

39. S. Ertekin, J. Huang, L. Bottou, and L. Giles, "Learning on the border: Active learning in imbalanced data classification," in *Proceedings of the 16th ACM Conference on Information and Knowledge Management* (Lisbon, Portugal), pp. 127–136, ACM, 2007.

40. R. Batuwita and V. Palade, "FSVM-CIL: Fuzzy support vector machines for class imbalance learning," *IEEE Transactions on Fuzzy Systems*, vol. 18, no. 3, pp. 558–571, 2010.

41. C.-F. Lin and S.-D. Wang, "Fuzzy support vector machines," *IEEE Transactions on Neural Networks*, vol. 13, no. 2, pp. 464–471, 2002.

42. C.-F. Lin and S.-D. Wang, "Training algorithms for fuzzy support vector machines with noisy data," *Pattern Recognition Letters*, vol. 25, no. 14, pp. 1647–1656, 2004.

43. P. Li, K. Chan, and W. Fang, "Hybrid kernel machine ensemble for imbalanced data sets," in *Proceedings of the 18th International Conference on Pattern Recognition*, pp. 1108–1111, IEEE Computer Society, 2006.

6

CLASS IMBALANCE AND ACTIVE LEARNING

Josh Attenberg

Etsy, Brooklyn, NY, USA and NYU Stern School of Business, New York, NY USA

Şeyda Ertekin

MIT Sloan School of Management, Massachusetts Institute of Technology, Cambridge, MA, USA

Abstract: The performance of a predictive model is tightly coupled with the data used during training. While using more examples in the training will often result in a better informed, more accurate model; limits on computer memory and real-world costs associated with gathering labeled examples often constrain the amount of data that can be used for training. In settings where the number of training examples is limited, it often becomes meaningful to carefully see just which examples are selected. In *active learning* (AL), the model itself plays a hands-on role in the selection of examples for labeling from a large pool of unlabeled examples. These examples are used for model training. Numerous studies have demonstrated, both empirically and theoretically, the benefits of AL: Given a fixed budget, a training system that interactively involves the current model in selecting the training examples can often result in a far greater accuracy than a system that simply selects random training examples. Imbalanced settings provide special opportunities and challenges for AL. For example, while AL can be used to build models that counteract the harmful effects of learning under class imbalance, extreme class imbalance can cause an AL strategy to "fail," preventing the selection scheme from choosing any useful examples for labeling. This chapter focuses on the interaction between AL and class imbalance, discussing (i) AL techniques designed specifically for dealing with imbalanced settings, (ii) strategies that leverage AL to overcome the deleterious effects of class imbalance, (iii) how extreme class imbalance can prevent AL systems from selecting useful examples, and alternatives to AL in these cases.

Imbalanced Learning: Foundations, Algorithms, and Applications, First Edition.
Edited by Haibo He and Yunqian Ma.
© 2013 The Institute of Electrical and Electronics Engineers, Inc. Published 2013 by John Wiley & Sons, Inc.

6.1   INTRODUCTION

The rich history of predictive modeling has been culminated in a diverse set of techniques capable of making accurate predictions on many real-world problems. Many of these techniques demand *training*, whereby a set of instances with ground-truth *labels* (values of a dependent variable) are observed by a model-building process that attempts to capture, at least in part, the relationship between the features of the instances and their labels. The resultant model can be applied to instances for which the label is not known, to estimate or predict the labels. These predictions depend not only on the functional structure of the model itself, but also on the particular data with which the model was trained. The accuracy of the predicted labels depends highly on the model's ability to capture an unbiased and sufficient understanding of the characteristics of different classes; in problems where the prevalence of classes is imbalanced, it is necessary to prevent the resultant model from being skewed toward the majority class and to ensure that the model is capable of reflecting the true nature of the minority class.

Another consequence of class imbalance is observed in domains where the ground-truth labels in the dataset are not available beforehand and need to be gathered on-demand at some cost. The costs associated with collecting labels may be due to human labor or is the result of costly incentives, interventions, or experiments. In these settings, simply labeling all available instances may not be practicable because of the budgetary constraints or simply a strong desire to be cost efficient. As in predictive modeling with imbalanced classes, the goal here is to ensure that the budget is not predominantly expended on getting the labels of the majority class instances, and to make sure that the set of instances to be labeled have comparable number of minority class instances as well.

In the context of learning from imbalanced datasets, the role of active learning (AL) can be viewed from two different perspectives. The first perspective considers the case where the labels for all the examples in a reasonably large, imbalanced dataset are readily available. The role of AL in this case is to reduce, and potentially eliminate, any adverse effects that the class imbalance can have on the model's generalization performance. The other perspective addresses the setting where we have prior knowledge that the dataset is imbalanced, and we would like to employ AL to select informative examples both from the majority and minority classes for labeling, subject to the constraints of a given budget. The first perspective focuses on AL's ability to address class imbalance, whereas the second perspective is concerned with the impact of class imbalance on the sampling performance of the active learner. The intent of this chapter is to present a comprehensive analysis of this interplay of AL and class imbalance. In particular, we first present techniques for dealing *with* the class imbalance problem using AL and discuss how AL can alleviate the issues that stem from class imbalance. We show that AL, even without any adjustments to target class imbalance, is an effective strategy to have a balanced view of the dataset in most cases. It is also possible to further improve the effectiveness of AL by tuning its sampling strategy in a class-specific way. Additionally, we will focus on dealing with highly skewed datasets and their impact on the selections performed by an AL strategy. Here, we discuss the impact significant class imbalance has on AL and illustrate alternatives to traditional AL that may be considered when dealing with the most difficult, highly skewed problems.

6.2 ACTIVE LEARNING FOR IMBALANCED PROBLEMS

The intent of this section is to provide the reader with some background on the AL problem in the context of building cost-effective classification models. We then discuss challenges encountered by AL heuristics in settings with significant class imbalance. We then discuss the strategies specialized in overcoming the difficulties imposed by this setting.

6.2.1 Background on Active Learning

AL is a specialized set of machine learning techniques developed for reducing the annotation costs associated with gathering the training data required for building predictive statistical models. In many applications, *unlabeled* data comes relatively cheaply when compared to the costs associated with the acquisition of a ground-truth value of the target variable of that data. For instance, the textual content of a particular web page may be crawled readily, or the actions of a user in a social network may be collected trivially by mining the web logs in that network. However, knowing with some degree of certainty the topical categorization of a particular web page, or identifying any malicious activity of a user in a social network is likely to require costly editorial review. These costs restrict the number of examples that may be labeled, typically to a small fraction of the overall population. Because of these practical constraints typically placed on the overall number of ground-truth labels available and the tight dependence of the performance of a predictive model on the examples in its training set, the benefits of careful selection of the examples are apparent. This importance is further evidenced by the vast research literature on the topic.

While an in-depth literature review is beyond the scope of this chapter, for context we provide a brief overview of some of the more broadly cited approaches in AL. For a more thorough treatment on the history and details of AL, we direct the reader to the excellent survey by Settles [1]. AL tends to focus on two scenarios—(i) stream-based selection, where unlabeled examples are presented one at a time to a predictive model, which feeds predicted target values to a consuming process and subsequently applies an AL heuristic to decide whether some budget should be expended gathering this example's class label for subsequent re-training. (ii) pool-based AL, on the other hand, is typically an offline, iterative process. Here, a large set of unlabeled examples are presented to an AL system. During each epoch of this process, the AL system chooses one or more unlabeled examples for labeling and subsequent model training. This proceeds until the budget is exhausted or some stopping criterion is met. At this time, if the

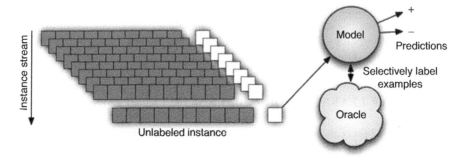

Figure 6.1 Pool-based active learning.

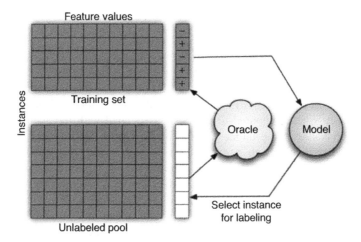

Figure 6.2 Stream-based active learning.

predictive performance is sufficient, the model may be incorporated into an end system that feeds the model with unlabeled examples and consumes the predicted results. A diagram illustrating these two types of AL scenarios is presented in Figures 6.1 and 6.2, respectively. Owing to the greater attention given to pool-based AL in recent scientific literature and the additional power available to an AL system capable of processing a large representation of the problem space at once, the remainder of this chapter focuses on the latter of these two scenarios, the pool-based setting.

The most common techniques in AL have focused on selecting examples from a so-called region of uncertainty, the area nearest to the current model's predictive decision boundary.[1] Incorporating uncertainty into active data acquisition dates back to research focused on optimal experimental design [2], and has been

---

[1]For instance, the simplest case when performing binary classification would involve choosing $x' = \arg\max_x \min_y P(y|x), y \in 0, 1$.

among the earliest successful examples of active machine learning techniques [3, 4]. The intuition behind uncertainty-based selection is that this region surrounding a model's decision boundary is where that model is most likely to make mistakes. Incorporating labeled examples from this region may improve the model's performance along this boundary, leading to gains in overall accuracy.

Many popular subsequent techniques are specializations of uncertainty selection, including query-by-committee-based approaches [5–7], where, given an ensemble of (valid) predictive models, examples are selected based on the level of disagreement elicited among the ensemble, and the popular "simple margin" technique proposed by Tong and Koller [8], where, given a current parameterization of a support vector machine (SVM), w_j, the example x_i is chosen that comes closest to the decision boundary, $x_i = \arg \min_{x'} |w_j \Phi(x)|$, where $\Phi(\cdot)$ is a function mapping an example to an alternate space utilized by the kernel function in the SVM: $k(u, v) = \Phi(u)\Phi(v)$.

Expected-utility-based approaches, where examples are chosen based on the estimated expected improvement in a certain objective, are achieved by incorporating a given example into the training set.[2] Such techniques often involve costly nested cross-validation where each available example is assigned all possible label states [9–11].

6.2.2 Dealing with the Class Imbalance Problem in Active Learning

Selecting examples from an unlabeled pool with substantial class imbalance may pose several difficulties for traditional AL. The greater proportion of examples in the majority class may lead to a model that prefers one class over another. If the labels of examples selected by an AL scheme are thought of as a random variable, the innate class imbalance in the example pool would almost certainly lead to a preference for majority examples in the training set. Unless properly dealt with,[3] this over-representation may simply lead to a predictive preference for the majority class when labeling. Typically, when making predictive models in an imbalanced setting, it is the minority class that is of interest. For instance, it is important to discover patients who have a rare but dangerous ailment based on the results of a blood test, or infrequent but costly fraud in a credit card company's transaction history. This difference in class preferences between an end system's needs and a model's tendencies causes a serious problem for AL (and predictive systems in general) in imbalanced settings. Even if the problem of highly imbalanced (although correct in terms of base rate) training set problem can be dealt with, the tendency for a selection algorithm to gather majority examples creates other problems. The nuances of the minority set may be poorly represented in the training data, leading to a "predictive misunderstanding" in

---

[2]Note that this selection objective may not necessarily be the same objective used during the base model's use time. For instance, examples may be selected according to their contribution to the reduction in problem uncertainty.
[3]For instance, by imbalanced learning techniques described throughout this book.

certain regions of the problem space; while a model may be able to accurately identify large regions of the minority space, portions of this space may get mislabeled, or labeled with poor quality, because of underrepresentation in the training set. At the extreme, disjunctive subregions may get missed entirely. Both of these problems are particularly acute as the class imbalance increases, and are discussed in greater detail in Section 6.6. Finally, in the initial stages of AL, when the base model is somewhat naïve, the minority class may get missed entirely as an AL heuristic probes the problem space for elusive but critical minority examples.

6.2.3   Addressing the Class Imbalance Problem with Active Learning

As we will demonstrate in Section 6.3, AL presents itself as an effective strategy for dealing with moderate class imbalance even without any special considerations for the skewed class distribution. However, the previously discussed difficulties imposed by more substantial class imbalance on the selective abilities of AL heuristics have led to the development of several techniques that have been specially adapted to imbalanced problem settings. These skew-specialized AL techniques incorporate an innate preference for the minority class, leading to more balanced training sets and better predictive performance in imbalanced settings. Additionally, there exists a category of density-sensitive AL techniques, techniques that explicitly incorporate the geometry of the problem space. By incorporating the knowledge of independent dimensions of the unlabeled example pool, there exists a potential for better exploration, resulting in improved resolution of rare subregions of the minority class. We detail these two broad classes of AL techniques as follows.

6.2.3.1   Density-Sensitive Active Learning   Utility-based selection strategies for AL attribute some score, $\mathcal{U}(\cdot)$, to each instance x encapsulating how much improvement can be expected from training on that instance. Typically, the examples offering a maximum $\mathcal{U}(x)$ are selected for labeling. However, the focus on individual examples may expose the selection heuristic to outliers, individual examples that achieve a high utility score, but do not represent a sizable portion of the problem space. Density-sensitive AL heuristics seek to alleviate this problem by leveraging the entire unlabeled pool of examples available to the active learner. By explicitly incorporating the geometry of the input space when attributing some selection score to a given example, outliers, noisy examples, and sparse areas of the problem space may be avoided. The following are some exemplary AL heuristics that leverage density sensitivity.

Information Density.   This is a general density-sensitive paradigm compatible with arbitrary utility-based active selection strategies and a variety of metrics used to compute similarity [12]. In this case, a meta-utility score is computed for each example based not only on a traditional utility score, $\mathcal{U}(x)$, but also on a measurement of that example's similarity to other instances in the problem space. Given a similarity metric between two points, $\text{sim}(x, x')$, information density selects examples according to:

$$\mathcal{U}_m(x) = \mathcal{U}(x) \left(\frac{1}{|X|} \sum_{x' \in X} \text{sim}(x, x') \right)^{\beta}$$

Here, β is a hyper-parameter controlling the trade-off between the raw instance-specific utility, $\mathcal{U}(x)$ and the similarity component in the overall selection criterion.

Zhu et al. [13] developed a technique similar to the information density technique of Settles and Craven, selecting the instances according a uncertainty-based criterion modified by a density factor: $\mathcal{U}_n(x) = \mathcal{U}(x)\text{KNN}(x)$, where $\text{KNN}(x)$ is the average cosine similarity of the K nearest neighbors to x. The same authors also propose the *sampling by clustering*, a density-only AL heuristic where the problem space is clustered, and the points closest to the cluster centeroids are selected for labeling.

Pre-Clustering. Here it is assumed that the problem is expressed as a mixture model comprising K distributions, each component model completely encoding information related to the labels of member examples—the label y is conditionally independent of the covariates x given knowledge of its cluster, k [14]. This assumption yields a joint distribution describing the problem: $p(x, y, k) = p(x|k)p(y|k)p(k)$, yielding a poster probability on y:

$$p_k(y|x) = \sum_{k=1}^{K} p(y|k) \frac{p(x|k)p(k)}{p(x)}$$

In essence, this a density-weighted mixture model used for classification. The K clusters are created by a application of typical clustering techniques of the data, with a cluster size used to estimate $p(k)$, and $p(y|k)$ is estimated via a logistic regression considering a cluster's representative example. A probability density is inferred for each cluster; in the example case presented in the earlier-mentioned work, an isotropic normal distribution is used, from which $p(x|k)$ can be estimated. Examples are then selected from an uncertainty score computed via the above-mentioned posterior model weighted by the probability of observing a given x:

$$\mathcal{U}_k(x) = (1 - |p_k(y|x)|)\, p(x)$$

Of course, there exists a variety of other techniques in the research literature designed to explicitly incorporate information related to the problem's density into an active selection criterion. McCallum and Nigam [15] modify a query-by-committee to use an exponentiated Kullback–Leibler (KL) divergence-based uncertainty metric and combine this with semi-supervised learning in the form of an expectation maximization (EM) procedure. This combined semi-supervised AL has the benefit of ignoring regions that can be reliably "filled in" by a semi-supervised procedure, while also selecting those examples that may benefit this EM process.

Donmez et al. [16] propose a modification of the density-weighted technique of Nguyen and Smeulders. This modification simply selects examples according to the convex combination of the density-weighted technique and traditional uncertainty sampling. This hybrid approach is again incorporated into a so-called dual-active learner, where only uncertainty sampling is incorporated once the benefits of pure density-sensitive sampling seem to be diminishing.

Alternate Density-Sensitive Heuristics. Donmez and Carbonell [17] incorporate density into active label selection by performing a change of coordinates into a space whose metric expresses not only Euclidian similarity but also density. Examples are then chosen based on a density-weighted uncertainty metric designed to select examples in pairs—one member of the pair from each side of the current decision boundary. The motivation is that sampling from both sides of the decision boundary may yield better results than selecting from one side in isolation.

Through selection based on an "unsupervised" heuristic estimating the utility of label acquisition on the pool of unlabeled instances, Roy and McCallum [9] incorporate the geometry of the problem space into active selection implicitly. This approach attempts to quantify the improvement in model performance attributable to each unlabeled example, taken in expectation over all label assignments:

$$\mathcal{U}_E = \sum_{y' \in Y} \hat{p}(y'|x) \sum_{x' \neq X} \mathcal{U}_e(x'; x, y = y')$$

Here, the probability of class membership in the earlier-mentioned expectation comes from the base model's current posterior estimates. The utility value on the right side of the previous equation, $\mathcal{U}_e(x'; x, y = y')$, comes from assuming a label of y', for example, x, and incorporating this pseudo-labeled example into the training set temporarily. The improvement in model performance with the inclusion of this new example is then measured. Since a selective label acquisition procedure may result in a small or arbitrarily biased set of examples, accurate evaluation through nested cross-validation is difficult. To accommodate this, Roy and McCallum propose two uncertainty measures taken over the pool of unlabeled examples, $x' \neq x$. Specifically, they look at the entropy of the posterior probabilities of examples in the pool, and the magnitude of the maximum posterior as utility measures, both estimated after the inclusion of the "new" example. Both metrics favor "sharp" posteriors, an optimization minimizing uncertainty rather than model performance; instances are selected by their reduction in uncertainty taken in expectation over the entire example pool.

6.2.3.2 Skew-Specialized Active Learning Additionally, there exists a body of research literature on AL specifically to deal with class imbalance problem. Tomanek and Hahn [18] investigates query-by-committee-based approaches to sampling labeled sentences for the task of named entity recognition. The goal of their selection strategy is to encourage class-balanced selections by incorporating class-specific costs. Unlabeled instances are ordered by a class-weighted, entropy-based disagreement measure, $-\sum_{j\in\{0,1\}} b_j V(k_j)/|C| \log V(k_j)/|C|$, where $V(k_j)$ is the number of votes from a committee of size $|C|$ that an instance belongs to a class k_j. b_j is a weight corresponding to the importance of including a certain class; a larger value of b_j corresponds to a increased tendency to include examples that are thought to belong to this class. From a window W of examples with highest disagreement, instances are selected greedily based on the model's estimated class membership probabilities so that the batch selected from the window has the highest probability of having a balanced class membership.

SVM-based AL has been shown [19] to be a highly effective strategy for addressing class imbalance without any skew-specific modifications to the algorithm. Bloodgood and Shanker [20] extend the benefits of SVM-based AL by proposing an approach that incorporates class-specific costs. That is, the typical C factor describing an SVM's misclassification penalty is broken up into C_+ and C_-, describing the costs associated with misclassification of positive and negative examples, respectively, a common approach for improving the performance of SVMs in cost-sensitive settings. Additionally, cost-sensitive SVMs are known to yield predictive advantages in imbalanced settings by offering some preference to an otherwise overlooked class, often using the heuristic for setting class-specific costs: $C_+/C_- = |\{x|x \in -\}|/|\{x|x \in +\}|$, a ratio in inverse proportion to the number of examples in each class. However, in the AL setting, the true class ratio is unknown, and the quantity C_+/C_- must be estimated by the AL system. Bloodgood and Shanker show that it is advantageous to use a preliminary stage of random selection in order to establish some estimate of the class ratio, and then proceed with example selection according to the uncertainty-based "simple margin" criterion using the appropriately tuned cost-sensitive SVM.

AL has also been studied as a way to improve the generalization performance of resampling strategies that address class imbalance. In these settings, AL is used to choose a set of instances for labeling, with sampling strategies used to improve the class distribution. Ertekin [21] presented virtual instance resampling technique using active learning (VIRTUAL), a hybrid method of oversampling and AL that forms an adaptive technique for resampling of the minority class instances. The learner selects the most informative example x_i for oversampling, and the algorithm creates a synthetic instance along the direction of x_i's one of k neighbors. The algorithm works in an online manner and builds the classifier incrementally without the need to retrain on the entire labeled dataset after creating a new synthetic example. This approach, which we present in detail in Section 6.4, yields an efficient and scalable learning framework.

Zhu and Hovy [22] describe a bootstrap-based oversampling strategy (BootOS) that, given an example to be resampled, generates a bootstrap example based on all the k neighbors of that example. At each epoch, the examples with the greatest uncertainty are selected for labeling and incorporated into a labeled set, L. From L, the proposed oversampling strategy is applied, yielding a more balanced dataset, L', a dataset that is used to retrain the base model. The selection of the examples with the highest uncertainty for labeling at each iteration involves resampling the labeled examples and training a new classifier with the resampled dataset; therefore, scalability of this approach may be a concern for large-scale datasets.

In the next section, we demonstrate that the principles of AL are naturally suited to address the class imbalance problem and that AL can in fact be an effective strategy to have a balanced view of an otherwise imbalanced dataset, without the need to resort to resampling techniques. It is worth noting that the goal of the next section is not to cast AL as a replacement for resampling strategies. Rather, our main goal is to demonstrate how AL can alleviate the issues that stem from class imbalance and present AL as an alternate technique that should be considered in case a resampling approach is impractical, inefficient, or ineffective. In problems where resampling *is* the preferred solution, we show in Section 6.4 that the benefits of AL can still be leveraged to address class imbalance. In particular, we present an adaptive oversampling technique that uses AL to determine which examples to resample in an online setting. These two different approaches show the versatility of AL and the importance of selective sampling to address the class imbalance problem.

6.3 ACTIVE LEARNING FOR IMBALANCED DATA CLASSIFICATION

As outlined in Section 6.2.1, AL is primarily considered as a technique to reduce the number of training samples that need to be labeled for a classification task. From a traditional perspective, the active learner has access to a vast pool of unlabeled examples, and it aims to make a clever choice to select the most informative example to obtain its label. However, even in the cases where the labels of training data are already available, AL can still be leveraged to obtain the informative examples through training sets [23–25]. For example, in large-margin classifiers such as SVM, the *informativeness* of an example is synonymous with its distance to the hyperplane. The farther an example is to the hyperplane, the more the learner is confident about its true class label; hence there is little, if any, benefit that the learner can gain by asking for the label of that example. On the other hand, the examples close to the hyperplane are the ones that yield the most information to the learner. Therefore, the most commonly used AL strategy in SVMs is to check the distance of each unlabeled example to the hyperplane and focus on the examples that lie closest to the hyperplane, as they are considered to be the most informative examples to the learner [8].

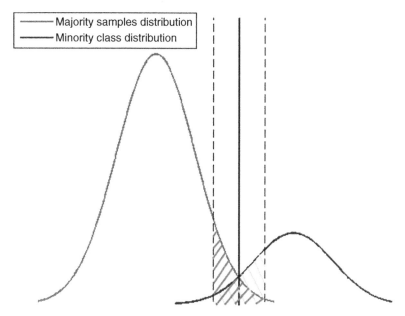

Figure 6.3   Data within the margin is less imbalanced than the entire data.

The strategy of selecting examples within the margin also strongly addresses the problems that arise from imbalanced classes. Consider the class distributions of an imbalanced dataset presented in Figure 6.3. The shaded region corresponds to the class distribution of the data within the margin. As shown in the figure, the imbalance ratio of the classes within the margin is much smaller than the class imbalance ratio of the entire dataset. Therefore, any selection strategy that focuses on the examples in the margin most likely ends up with a more balanced class distribution than that of the entire dataset.

Throughout this section, the discussion is constrained to standard two-class classification problems using SVMs. The next section presents a brief overview of SVMs, followed by the working principles of an efficient AL algorithm in Section 6.3.2. We explain the advantage of using online SVMs with the active sample selection in Section 6.3.3.

6.3.1   Support Vector Machines

SVMs [26] are well known for their strong theoretical foundations, generalization performance, and ability to handle high dimensional data. In the binary classification setting, let $((x_1, y_1) \cdots (x_n, y_n))$ be the training dataset, where x_i are the feature vectors representing the instances and $y_i \in (-1, +1)$ be the labels of the instances. Using the training set, SVM builds an optimum hyperplane—a linear discriminant in a higher dimensional feature space—that separates the two classes by the largest margin. This hyperplane is obtained by minimizing the following objective function:

$$\min_{\mathbf{w},b,\xi_i} \frac{1}{2}\mathbf{w} \cdot \mathbf{w}^{\mathrm{T}} + C \sum_{i=1}^{N} \xi_i \tag{6.1}$$

$$\text{subject to} \begin{cases} \forall i \ y_i(\mathbf{w}^{\mathrm{T}}\Phi(x_i) - b) \geq 1 - \xi_i \\ \forall i \ \xi_i \geq 0 \end{cases} \tag{6.2}$$

where $\mathbf{w}$ is the norm of the hyperplane, b is the offset, y_i are the labels, $\Phi(\cdot)$ is the mapping from input space to feature space, and ξ_i are the slack variables that permit the non-separable case by allowing misclassification of training instances. In practice, the convex quadratic programming (QP) problem in Equation 6.1 is solved by optimizing the dual cost function. The dual representation of Equation 6.1 is given as

$$\max W(\alpha) \equiv \sum_{i=1}^{N} \alpha_i - \frac{1}{2} \sum_{i,j} \alpha_i \alpha_j y_i y_j K(\mathbf{x_i}, \mathbf{x_j}) \tag{6.3}$$

$$\text{subject to} \begin{cases} \forall i \ 0 \leq \alpha_i \leq C \\ \sum_{i=1}^{N} \alpha_i y_i = 0 \end{cases} \tag{6.4}$$

where y_i are the labels, $\Phi(\cdot)$ is the mapping from the input space to the feature space, $K(\mathbf{x_i}, \mathbf{x_j}) = \langle \Phi(\mathbf{x_i}), \Phi(\mathbf{x_j}) \rangle$ is the kernel matrix, and the α_i's are the *Lagrange multipliers*, which are nonzero only for the training instances that fall in the margin. Those training instances are called *support vectors* and they define the position of the hyperplane. After solving the QP problem, the norm of the hyperplane $\mathbf{w}$ can be represented as

$$\mathbf{w} = \sum_{i=1}^{n} \alpha_i \Phi(x_i) \tag{6.5}$$

6.3.2 Margin-Based Active Learning with SVMs

Note that in Equation 6.5, only the support vectors affect the SVM solution. This means that if SVM is retrained with a new set of data that consist of only those support vectors, the learner will end up finding the same hyperplane. This emphasizes the fact that not all examples are equally important in training sets. Then the question becomes how to select the most informative examples for labeling from the set of unlabeled training examples. This section focuses on a form of selection strategy called *margin-based AL*. As was highlighted earlier, in SVMs, the most informative example is believed to be the closest one to the hyperplane as it divides the *version space* into two equal parts. The aim is to reduce the version space as fast as possible to reach the solution faster in order to avoid certain *costs* associated with the problem. For the possibility of a nonsymmetric version space, there are more complex selection methods suggested by Tong and Koller [8], but the advantage of those methods are not significant, considering their high computational costs.

6.3.2.1 Active Learning with Small Pools The basic working principle of margin-based AL with SVMs is: (i) train an SVM on the existing training data, (ii) select the closest example to the hyperplane, and (iii) add the new selected example to the training set and train again. In classical AL [8], the search for the most informative example is performed over the entire dataset. Note that, each iteration of AL involves the recomputation of each training example's distance to the new hyperplane. Therefore, for large datasets, searching the entire training set is a very time-consuming and computationally expensive task.

One possible remedy for this performance bottleneck is to use the "59 trick" [27], which alleviates a full search through the entire dataset, approximating the most informative examples by examining a small constant number of randomly chosen samples. The method picks L ($L \ll$ # training examples) random training samples in each iteration and selects the best (closest to the hyperplane) among them. Suppose, instead of picking the closest example among all the training samples $X_N = (x_1, x_2, \ldots, x_N)$ at each iteration, we first pick a random subset X_L, $L \ll N$ and select the closest sample x_i from X_L based on the condition that x_i is among the top $p\%$ closest instances in X_N with probability $(1 - \eta)$. Any numerical modification to these constraints can be met by varying the size of L, and is independent of N. To demonstrate this, the probability that at least one of the L instances is among the closest p is $1 - (1 - p)^L$. Owing to the requirement of $(1 - \eta)$ probability, we have

$$1 - (1 - p)^L = 1 - \eta \qquad (6.6)$$

which follows the solution of L in terms of η and p

$$L = \frac{\log \eta}{\log(1 - p)} \qquad (6.7)$$

For example, the active learner will pick one example, with 95% probability, that is among the top 5% closest instances to the hyperplane, by randomly sampling only $\lceil \log(0.05) / \log(0.95) \rceil = 59$ examples regardless of the training set size. This approach scales well since the size of the subset L is independent of the training set size N, requires significantly less training time, and does not have an adverse effect on the classification performance of the learner.

6.3.3 Active Learning with Online Learning

Online learning algorithms are usually associated with problems where the complete training set is not available. However, in cases where the complete training set *is* available, the computational properties of these algorithms can be leveraged for faster classification and incremental learning. Online learning techniques can process new data presented one at a time, as a result of either AL or random selection, and can integrate the information of the new data to the system without training on all previously seen data, thereby allowing models to be constructed incrementally. This working principle of online learning algorithms leads to speed improvements and a reduced memory footprint, making the algorithm applicable to very large datasets. More importantly, this incremental learning principle suits the nature of AL much more naturally than the batch algorithms. Empirical evidence indicates that a single presentation of each training example to the algorithm is sufficient to achieve training errors comparable to those achieved by the best minimization of the SVM objective [24].

6.3.4   Performance Metrics

Classification accuracy is not a good metric to evaluate classifiers in applications facing class imbalance problems. SVMs have to achieve a trade-off between maximizing the margin and minimizing the empirical error. In the non-separable case, if the misclassification penalty C is very small, the SVM learner simply tends to classify every example as negative. This extreme approach maximizes the *margin* while making no classification errors on the negative instances. The only error is the cumulative error of the positive instances that are already few in numbers. Considering an imbalance ratio of 99 to 1, a classifier that classifies everything as negative, will be 99% accurate. Obviously, such a scheme would not have any practical use, as it would be unable to identify positive instances.

For the evaluation of these results, it is useful to consider several other prediction performance metrics such as g-means, area under the curve (AUC), and precision–recall break-even point (PRBEP), which are commonly used in imbalanced data classification. g-Means [28] is denoted as $g = \sqrt{\text{sensitivity} \cdot \text{specificity}}$, where sensitivity is the accuracy on the positive instances given as TruePos./(TruePos. + FalseNeg.), and specificity is the accuracy on the negative instances given as TrueNeg./(TrueNeg. + FalsePos.).

The receiver operating curve (ROC) displays the relationship between sensitivity and specificity at all possible thresholds for a binary classification scoring model, when applied to independent test data. In other words, ROC curve is a plot of the true positive rate against the false positive rate as the decision threshold is changed. The *area under the ROC* (AUROC or AUC) is a numerical measure of a model's discrimination performance and shows how successfully and correctly the model ranks and thereby separates the positive and negative observations. Since the AUC metric evaluates the classifier across the entire range of decision thresholds, it gives a good overview of the performance when the operating condition for the classifier is unknown or the classifier is expected to be used in situations with significantly different class distributions.

PRBEP is another commonly used performance metric for imbalanced data classification. PRBEP is the accuracy of the positive class at the threshold where precision equals to recall. Precision is defined as TruePos./(TruePos. + FalsePos.), and recall is defined as TruePos./(TruePos. + FalseNeg.)

6.3.5 Experiments and Empirical Evaluation

We study the performance of the algorithm on various benchmark real-world datasets, including MNIST, USPS, several categories of Reuters-21578 collection, five topics from CiteSeer, and three datasets from the University of California, Irvine (UCI) repository. The characteristics of the datasets are outlined in [19]. In the experiments, an early stopping heuristic for AL is employed, as it has been shown that AL converges to the solution faster than the random sample selection method [19]. A theoretically sound method to stop training is when the examples in the margin are exhausted. To check whether there are still unseen training examples in the margin, the distance of the newly selected example is compared against the support vectors of the current model. If the newly selected example by AL (closest to the hyperplane) is not closer than any of the support vectors, it is concluded that the margin is exhausted. A practical implementation of this idea is to count the number of support vectors during the AL training process. If the number of the support vectors stabilizes, it implies that all possible support vectors have been selected by the AL method.

As the first experiment, examples are randomly removed from the minority class in *Adult* dataset to achieve different data imbalance ratios, comparing SVM-based AL and random sampling (RS).[4] For brevity, AL with small pools is referred to as AL as the small pools heuristic is utilized for all AL methods considered later. Comparisons of PRBEP in Figure 6.4 show an interesting behavior. As the class imbalance ratio is increased, AL curves display peaks in the early steps of the learning. This implies that by using an early stopping criteria, AL can give higher prediction performance than RS can possibly achieve even after using all the training data. The learning curves presented in Figure 6.4 demonstrate that the addition of instances to a model's training after finding those most informative instances can be detrimental to the prediction performance of the classifier, as this may cause the model to suffer from overfitting. Figure 6.4 curves show that generalization can peak to a level above that can be achieved by using all available training data. In other words, it is possible to achieve better classification performance from a small informative subset of the training data than what can be achieved using all available training data. This finding agrees with that of Schohn and Cohn [23] and strengthens the idea of applying an early stopping to AL algorithms.

For further comparison of the performance of a model built on all available data (batch) and AL subject to early halting criteria, refer to Table 6.1, comparing the g-means and the AUC values for these two methods. The data efficiency column for AL indicates that by processing only a portion of the examples from the

---

[4]Here, the random process is assumed to be uniform; examples are selected with equal probability from the available pool.

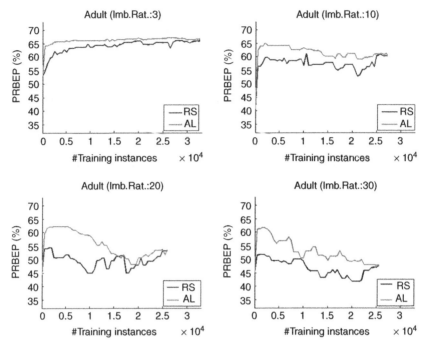

Figure 6.4 Comparison of PRBEP of AL and RS on the adult datasets with different imbalance ratios (Imb.R.=3, 10, 20, 30).

training set, AL can achieve similar or even higher generalization performance than that of batch, which sees all the training examples. Another important observation from Table 6.1 is that support vector imbalance ratios in the final models are much less than the class imbalance ratios of the datasets. This confirms the discussion of Figure 6.3. The class imbalance ratio within the margins is much less than that of the entire data, and AL can be used to reach those informative examples that most likely become support vectors without seeing all the training examples.

Figure 6.5 investigates how the number of support vectors changes when presented with examples selected according to AL and RS. Because the base rate of the dataset gathered by RS approaches that of the example pool, the support vector imbalance ratio quickly approaches the data imbalance ratio. As learning continues, the learner should gradually see all the instances within the final margin and the support vector imbalance ratio decreases. At the end of training with RS, the support vector imbalance ratio is the data imbalance ratio within the margin. The support vector imbalance ratio curve of AL is drastically different than RS. AL intelligently picks the instances closest to the margin in each step. Since the data imbalance ratio within the margin is lower than data imbalance ratio, the support vectors in AL are more balanced than RS during learning. Using AL, the model saturates by seeing only 2000 (among 7770) training instances and

Table 6.1 Comparison of *g*-Means and AUC for AL and RS with Entire Training Data (Batch)

	Dataset	g-Means (%) Batch	g-Means (%) AL	AUC (%) Batch	AUC (%) AL	Imb. Rat.	SV- / SV+	Data Efficiency (%)
Reuters	Corn	85.55	86.59	99.95	99.95	41.9	3.13	11.6
	Crude	88.34	89.51	99.74	99.74	19.0	2.64	22.6
	Grain	91.56	91.56	99.91	99.91	16.9	3.08	29.6
	Interest	78.45	78.46	99.01	99.04	21.4	2.19	30.9
	Money-fx	81.43	82.79	98.69	98.71	13.4	2.19	18.7
	Ship	75.66	74.92	99.79	99.80	38.4	4.28	20.6
	Trade	82.52	82.52	99.23	99.26	20.1	2.22	15.4
	Wheat	89.54	89.55	99.64	99.69	35.7	3.38	11.6
CiteSeer	AI	87.83	88.58	94.82	94.69	4.3	1.85	33.4
	COMM	93.02	93.65	98.13	98.18	4.2	2.47	21.3
	CRYPT	98.75	98.87	99.95	99.95	11.0	2.58	15.2
	DB	92.39	92.39	98.28	98.46	7.1	2.50	18.2
	OS	91.95	92.03	98.27	98.20	24.2	3.52	36.1
UCI	Abalone-7	100.0	100.0	100.0	100.0	9.7	1.38	24.0
	Letter-A	99.28	99.54	99.99	99.99	24.4	1.46	27.8
	Satimage	82.41	83.30	95.13	95.75	9.7	2.62	41.7
	USPS	99.22	99.25	99.98	99.98	4.9	1.50	6.8
	MNIST-8	98.47	98.37	99.97	99.97	9.3	1.59	11.7

SV ratios are given at the saturation point. Data efficiency corresponds to the percentage of training instances that AL processes to reach saturation.

crude7770

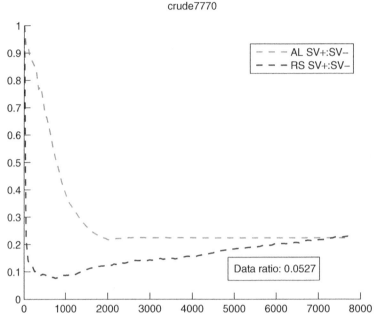

Figure 6.5 Support vector ratios in AL and RS.

reaches the final support vector imbalance ratio. Note that both methods achieve similar support vector imbalance ratios when learning finishes, but AL achieves this in the early steps of the learning.

It is also interesting to consider the performance of AL as a selection heuristic in light of more conventional sampling strategies. Here, AL is compared to traditional under-sampling of the majority class (US) and an oversampling method (SMOTE, synthetic minority oversampling technique), both being examples of resampling techniques that require preprocessing. It has been shown that oversampling at random does not help to improve prediction performance [29]; therefore, a more complex oversampling method is required. SMOTE oversamples the minority class by creating synthetic examples rather than with replacement. The k nearest positive neighbors of all positive instances are identified, and synthetic positive examples are created and placed randomly along the line segments joining the k-minority class nearest neighbors.

For additional comparison, the method of assigning different costs (DCs) to the positive and negative classes as the misclassification penalty parameter is examined. For instance, if the imbalance ratio of the data is $19 : 1$ in favor of the negative class, the cost of misclassifying a positive instance is set to be 19 times greater than that of misclassifying a negative one. We use the online SVM package LASVM[5] in all experiments. Other than the results of the methods

---

[5]Available at http://leon.bottou.org/projects/lasvm

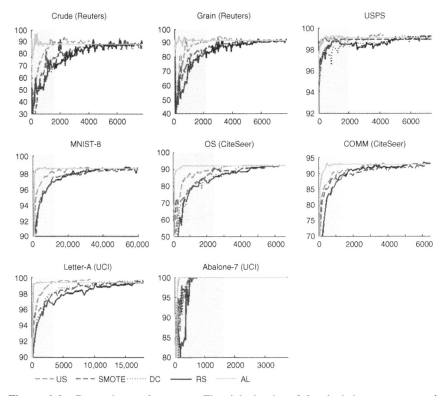

Figure 6.6 Comparisons of g-means. The right border of the shaded area corresponds to the early stopping point.

addressing the class imbalance problem, the results of batch algorithm with the original training set are provided to form a baseline. LASVM is run in RS mode for US, SMOTE, and DC.

We present the comparisons of the methods for g-means performance metric for several datasets in Figure 6.6. The right border of the shaded light gray area is the place where the aforementioned early stopping strategy is applied. The curves in the graphs are averages of 10 runs. For completeness, all AL experiments were allowed to continue to select examples until exhaustion, bypassing any early stopping. Table 6.2 presents the PRBEP of the methods and the total running times of the SMOTE and AL on 18 benchmark and real-world datasets. The results for AL in Table 6.2 depict the results in the early stopping points. The results for the other methods in Table 6.2 depict the values at the end of the curves—when trained with the entire dataset—as those methods do not employ any early stopping criteria. We did not apply early stopping criteria to the other methods because, as observed from Figure 6.6, no early stopping criteria would achieve a comparable training time to that of AL's training time without a significant loss in their prediction performance based on convergence time. The other

Table 6.2 Comparison of PRBEP and Training Time

Metric		PRBEP					Training Time (s)	
	Dataset	Batch	US	SMOTE	DC	AL	SMOTE	AL
Reuters	Corn	91.07	78.57	91.07	89.28	89.29	87	16
	Crude	87.83	85.70	87.83	87.83	87.83	129	41
	Grain	92.62	89.93	91.44	91.94	91.94	205	50
	Interest	76.33	74.04	77.86	75.57	75.57	116	42
	Money-fx	73.74	74.30	75.42	75.42	76.54	331	35
	Ship	86.52	86.50	88.76	89.89	89.89	49	32
	Trade	77.77	76.92	77.77	77.78	78.63	215	38
	Wheat	84.51	81.61	84.51	84.51	85.92	54	25
CiteSeer	AI	78.80	80.68	78.99	78.79	79.17	1402	125
	COMM	86.59	86.76	86.59	86.59	86.77	1707	75
	CRYPT	97.89	97.47	97.89	97.89	97.89	310	19
	DB	86.36	86.61	86.98	86.36	86.36	526	41
	OS	84.07	83.19	84.07	84.07	84.07	93	23
UCI	Abalone-7	100.0	100.0	100.0	100.0	100.0	16	4
	Letter-A	99.48	96.45	99.24	99.35	99.35	86	3
	Satimage	73.46	68.72	73.46	73.93	73.93	63	21
	USPS	98.44	98.44	98.13	98.44	98.75	4328	13
	MNIST-8	97.63	97.02	97.74	97.63	97.74	83,339	1048

Table 6.3 Support Vectors with SMOTE (SMT), AL, and VIRTUAL

	Dataset	Imb. Rt.	#SV(-)/#SV(+)			#SV_V(+)/#V.I.	
			SMT	AL	VIRTUAL	SMT	VIRTUAL
Reuters	acq	3.7	1.24	1.28	1.18	2.4%	**20.3%**
	corn	41.9	2.29	3.08	1.95	17.1%	**36.6%**
	crude	19.0	2.30	2.68	2.00	10.8%	**50.4%**
	earn	1.7	1.68	1.89	1.67	6.0%	**24.2%**
	grain	16.9	2.62	3.06	2.32	7.2%	**42.3%**
	interest	21.4	1.84	2.16	1.66	13.3%	**72.2%**
	money-fx	13.4	1.86	2.17	1.34	8.2%	**31.1%**
	ship	38.4	3.45	4.48	2.80	20.0%	**66.5%**
	trade	20.1	1.89	2.26	1.72	15.4%	**26.6%**
	wheat	35.7	2.55	3.43	2.22	12.3%	**63.9%**
UCI	Abalone	9.7	0.99	1.24	0.99	30.4%	**69.2%**
	Breast	1.9	1.23	0.60	0.64	2.9%	**39.5%**
	Letter	24.4	1.21	1.48	0.97	0.98%	**74.4%**
	Satimage	9.7	1.31	1.93	0.92	37.3%	**53.8%**

Imb.Rt. is the data imbalance ratio, and #SV($-$)/#SV(+) represents the support vector imbalance ratio. The rightmost two columns compare the portion of the virtual instances selected as support vectors in SMOTE and VIRTUAL.

methods converge to similar levels of g-means when nearly all training instances are used, and applying an early stopping criteria would have little, if any, effect on their training times.

Since AL involves discarding some instances from the training set, it can be perceived as a type of under-sampling method. Unlike traditional US, which discards majority samples randomly, AL performs an intelligent search for the most informative ones adaptively in each iteration according to the current hyperplane. Datasets where class imbalance ratio is high such as *corn, wheat, letter*, and *satimage* observe significant decrease in PRBEP of US (Table 6.3). Note that US's under-sampling rate for the majority class in each category is set to the same value as the final support vector ratio where AL reaches in the early stopping point and RS reaches when it sees the entire training data. Although the class imbalance ratios provided to the learner in AL and US are the same, AL achieves significantly better PRBEP performance metric than US. The Wilcoxon-signed-rank test (two-tailed) reveals that the zero median hypothesis can be rejected at the significance level 1% ($p = 0.0015$), implying that AL performs statistically better than US in these 18 datasets. These results reveal the importance of using the informative instances for learning.

Table 6.2 gives the comparison of the computation times of the AL and SMOTE. Note that SMOTE requires significantly long preprocessing time that dominates the training time in large datasets, for example, MNIST-8 dataset. The low computation cost, scalability, and high prediction performance of AL suggest that AL can efficiently handle the class imbalance problem.

6.4 ADAPTIVE RESAMPLING WITH ACTIVE LEARNING

The analysis in Section 6.3.5 shows the effectiveness of AL on imbalanced datasets without employing any resampling techniques. This section extends the discussion on the effectiveness of AL for imbalanced data classification and demonstrates that even in cases where resampling is the preferred approach, AL can still be used to significantly improve the classification performance.

In supervised learning, a common strategy to overcome the rarity problem is to resample the original dataset to decrease the overall level of class imbalance. Resampling is done either by oversampling the minority (positive) class and/or under-sampling the majority (negative) class until the classes are approximately equally represented [28, 30–32]. Oversampling, in its simplest form, achieves a more balanced class distribution either by duplicating minority class instances or introducing new synthetic instances that belong to the minority class [30]. No information is lost in oversampling as all original instances of the minority and the majority classes are retained in the oversampled dataset. The other strategy to reduce the class imbalance is under-sampling, which eliminates some majority class instances mostly by RS.

Even though both approaches address the class imbalance problem, they also suffer some drawbacks. The under-sampling strategy can potentially sacrifice the prediction performance of the model, as it is possible to discard informative instances that the learner might benefit. Oversampling strategy, on the other hand, can be computationally overwhelming in cases with large training sets—if a complex oversampling method is used; a large computational effort must be expended during preprocessing of the data. Worse, oversampling causes longer training time during the learning process because of the increased number of training instances. In addition to suffering from increased runtime due to added computational complexity, it also necessitates an increased memory footprint due to the extra storage requirements of artificial instances. Other costs associated with the learning process (i.e., extended kernel matrix in kernel classification algorithms) further increase the burden of oversampling.

6.4.1 VIRTUAL: Virtual Instance Resampling Technique Using Active Learning

In this section, the focus is on the oversampling strategy for imbalanced data classification and investigate how it can benefit from the principles of AL. Our goal is to remedy the efficiency drawbacks of oversampling in imbalanced data classification and use an AL strategy to generate minority class instances only if they can be useful to the learner. VIRTUAL (virtual instance resampling technique using active learning) [22] is a hybrid method of oversampling and AL that forms an adaptive technique for resampling of the minority class instances. In contrast to traditional oversampling techniques that act as an *offline* step that generates virtual instances of the minority class before the training process, VIRTUAL leverages the power of AL to intelligently and adaptively oversample the data *during* training, removing the need for an offline and separate preprocessing stage. Similar to the discussions in the previous section, VIRTUAL also employs an online SVM-based AL strategy. In this setting, the informativeness of instances is measured by their distance to their hyperplane, and the most informative instances are selected as the support vectors. VIRTUAL targets the set of support vectors during training, and resamples new instances based on this set. Since most support vectors are found during early stages of training, corresponding virtual examples are also created in the early stages. This prevents the algorithm from creating excessive and redundant virtual instances, and integrating the resampling process into the training stage improves the efficiency and generalization performance of the learner compared to other competitive oversampling techniques.

6.4.1.1 Active Selection of Instances Let S denote the pool of real and virtual training examples unseen by the learner at each AL step. Instead of searching for the most informative instance among all the samples in S, VIRTUAL employs the small-pool AL strategy that is discussed in section 6.3.2. From the small pool, VIRTUAL selects an instance that is closest to the hyperplane according to the current model. If the selected instance is a real positive instance (from the original training data) and becomes a support vector, VIRTUAL advances to the oversampling step, explained in the following section. Otherwise, the algorithm proceeds to the next iteration to select another instance.

6.4.1.2 Virtual Instance Generation VIRTUAL oversamples the real minority instances (instances selected from the minority class of the original training data) that become support vectors in the current iteration. It selects the k nearest minority class neighbors $(x_{i \to 1} \cdots x_{i \to k})$ of x_i based on their similarities in the kernel-transformed higher dimensional feature space. We limit the neighboring instances of x_i to the minority class so that the new virtual instances lie within the minority class distribution. Depending on the amount of oversampling required, the algorithm creates v virtual instances. Each virtual instance lies on any of the line segments joining x_i and its neighbor $x_{i \to j}$ ($j = 1, \ldots, k$). In other words, a neighbor $x_{i \to j}$ is randomly picked and the virtual instance is created as $\overline{x}_v = \lambda \cdot x_i + (1 - \lambda)x_{i \to j}$, where $\lambda \in (0, 1)$ determines the placement of $\overline{x}_v$ between x_i and $x_{i \to j}$. All v virtual instances are added to S and are eligible to be picked by the active learner in the subsequent iterations.

The pseudocode of VIRTUAL given in Algorithm 6.1 depicts the two processes described previously. In the beginning, the pool S contains all real instances in the training set. At the end of each iteration, the instance selected is removed from S, and any virtual instances generated are included in the pool S. In this pseudocode, VIRTUAL terminates when there are no instances in S.

6.4.2 Remarks on VIRTUAL

We compare VIRTUAL with a popular oversampling technique SMOTE. Figure 6.7a shows the different behaviors of how SMOTE and VIRTUAL create virtual

Algorithm 6.1   VIRTUAL

Define:
$X = \{x_1, x_2, \cdots, x_n\}$: training instances
X_R^+ : positive real training instances
S : pool of training instances for SVM
v : # virtual instances to create in each iteration
L : size of the small set of randomly picked samples
      for active sample selection

---

1. Initialize $S \leftarrow X$
2. **while** $S \neq \emptyset$
3.   // *Active sample selection step*
4. $d_{min} \leftarrow \infty$
5.   **for** $i \leftarrow 1$ to L
6. $x_j \leftarrow RandomSelect(S)$
7.   **If** $d(x_j, hyperplane) < d_{min}$
8. $d_{min} \leftarrow d(x_j, hyperplane)$
9. $candidate \leftarrow x_j$
10.   **end**
11.   **end**
12. $x_s \leftarrow candidate$
13.   // *Virtual Instance Generation*
14.   **If** x_s becomes SV **and** $x_s \in X_R^+$
15. $K \leftarrow k$ nearest neighbors of x_s
16.   **for** $i \leftarrow 1$ to v
17. $x_m \leftarrow RandomSelect(K)$
18.   // Create a virtual positive instance $x_{s,m}^v$ between x_s and x_m
19. λ=random number between 0 and 1
20. $x_{s,m}^v = \lambda \cdot x_s + (1 - \lambda)x_m$
21. $S \leftarrow S \cup x_{s,m}^v$
22.   **end**
23.   **end**
24. $S \leftarrow S - x_s$
25. **end**

---

instances for the minority class. SMOTE creates virtual instance(s) for each positive example (Figure 6.7b), whereas VIRTUAL creates the majority of virtual instances around the positive canonical hyperplane (shown with a dashed line in Figure 6.7). Note that a large portion of virtual instances created by SMOTE is far away from the hyperplane and thus are not likely to be selected as support vectors. VIRTUAL, on the other hand, generates virtual instances near the real positive support vectors adaptively in the learning process. Hence, the virtual instances are near the hyperplane and thus are more informative.

We further analyze the computation complexity of SMOTE and VIRTUAL. The computation complexity of VIRTUAL is $O(|SV(+)| \cdot v \cdot C)$, where v is the

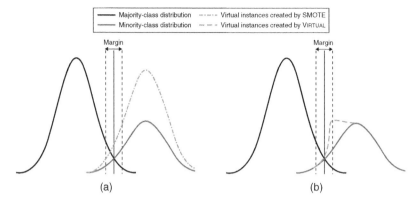

Figure 6.7 Comparison of oversampling the minority class with SMOTE and VIRTUAL. (a) Oversampling with SMOTE and (b) oversampling with VIRTUAL.

number of virtual instances created for a real positive support vector in each iteration, $|SV(+)|$ is the number of positive support vectors, and $\mathcal{C}$ is the cost of finding k nearest neighbors. The computation complexity of SMOTE is $O(|X_R^+| \cdot v \cdot \mathcal{C})$, where $|X_R^+|$ is the number of positive training instances. $\mathcal{C}$ depends on the approach for finding k nearest neighbors. The naive implementation searches all N training instances for the nearest neighbors and thus $\mathcal{C} = kN$. Using advanced data structure such as kd-tree, $\mathcal{C} = k \log N$. Since $|SV(+)|$ is typically much less than $|X_R^+|$, VIRTUAL incurs lower computation overhead than SMOTE. Also, with fewer virtual instances created, the learner is less burdened with VIRTUAL. We demonstrate with empirical results that the virtual instances created with VIRTUAL are more informative and the prediction performance is also improved.

6.4.3 Experiments

We conduct a series of experiments on Reuters-21578 and four UCI datasets to demonstrate the efficacy of VIRTUAL. The characteristics of the datasets are detailed in [22]. We compare VIRTUAL with two systems, AL and SMOTE. AL adopts the traditional AL strategy without preprocessing or creating any virtual instances during learning. SMOTE, on the other hand, preprocesses the data by creating virtual instances before training and uses RS in learning. Experiments elicit the advantages of adaptive virtual sample creation in VIRTUAL.

Figures 6.8 and 6.9 provide details on the behavior of the three algorithms, SMOTE, AL, and VIRTUAL. For the Reuters datasets (Fig. 6.9), note that in all the 10 categories, VIRTUAL outperforms AL in g-means metric after saturation. The difference in performance is most pronounced in the more imbalanced categories, for example, *corn*, *interest*, and *ship*. In the less imbalanced datasets such as *acq* and *earn*, the difference in g-means of both methods is less noticeable. The g-means of SMOTE converges much slower than both AL and VIRTUAL.

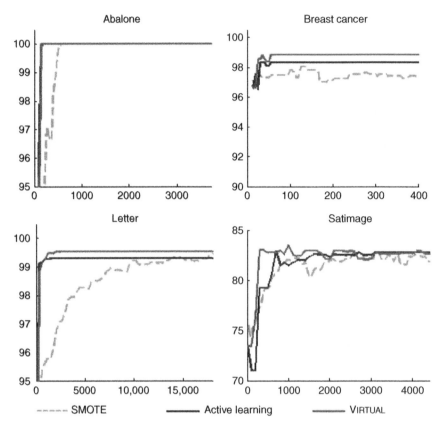

Figure 6.8 Comparison of SMOTE, AL, and VIRTUAL on *UCI* datasets. We present the *g*-means (%) (*y*-axis) of the current model for the test set versus the number of training samples (*x*-axis) seen.

However, SMOTE converges to higher *g*-means than AL in some of the categories, indicating that the virtual positive examples provide additional information that can be used to improve the model. VIRTUAL converges to the same or even higher *g*-means than SMOTE while generating fewer virtual instances. For the UCI datasets (Fig. 6.8), VIRTUAL performs as well as AL in *abalone* in *g*-means and consistently outperforms AL and SMOTE in the other three datasets.

In Table 6.4, the support vector imbalance ratios of all the three methods are lower than the data imbalance ratio, and VIRTUAL achieves the most balanced ratios of positive and negative support vectors in the Reuters datasets. Despite that the datasets used have different data distributions, the portion of virtual instances become support vectors in VIRTUAL consistently and significantly higher than that in SMOTE. These results confirm the previous discussion that VIRTUAL is more effective in generating informative virtual instances.

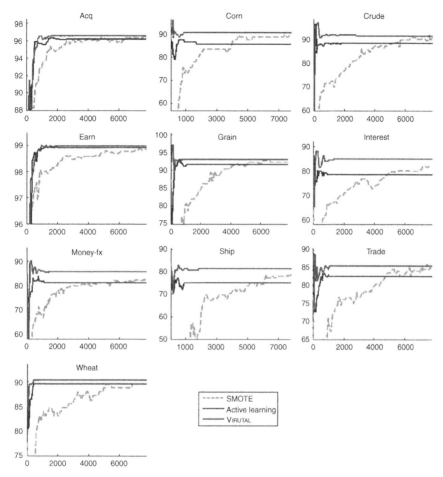

Figure 6.9 Comparison of SMOTE, AL, and VIRTUAL on 10 largest categories of *Reuters-21578*. We show the g-means (%) (y-axis) of the current model for the test set versus the number of training samples (x-axis) seen.

Table 6.4 presents the g-means and the total learning time for SMOTE, AL, and VIRTUAL. Classical batch SVM's g-means values are also provided as a reference point. In Reuters datasets, VIRTUAL yields the highest g-means in all categories. Table 6.4 shows the effectiveness of adaptive virtual instance generation. In categories *corn*, *interest*, and *ship* with high class imbalance ratio, VIRTUAL gains substantial improvement in g-means. Compared to AL, VIRTUAL requires additional time for the creation of virtual instances and selection of those that may become support vectors. Despite this overhead, VIRTUAL's training times are comparable with those of AL. In the cases where minority examples are abundant, SMOTE demands substantially longer time to create virtual instances than VIRTUAL. But as the rightmost columns in Table 6.3 show, only a small fraction

Table 6.4 g-Means and Total Learning Time Using SMOTE, AL, and VIRTUAL

		g-Means (%)				Total Learning time (s)		
	Dataset	Batch	SMOTE	AL	VIRTUAL	SMOTE	AL	VIRTUAL
Reuters	acq	96.19 (3)	96.21 (2)	96.19 (3)	**96.54 (1)**	2271	146	203
	corn	85.55 (4)	89.62 (2)	86.59 (3)	**90.60 (1)**	74	43	66
	crude	88.34 (4)	91.21 (2)	88.35 (3)	**91.74 (1)**	238	113	129
	earn	98.92 (3)	**98.97 (1)**	98.92 (3)	**98.97 (1)**	4082	121	163
	grain	91.56 (4)	92.29 (2)	91.56 (4)	**93.00 (1)**	296	134	143
	interest	78.45 (4)	83.96 (2)	78.45 (4)	**84.75 (1)**	192	153	178
	money-fx	81.43 (3)	83.70 (2)	81.08 (4)	**85.61 (1)**	363	93	116
	ship	75.66 (3)	78.55 (2)	74.92 (4)	**81.34 (1)**	88	75	76
	trade	82.52 (3)	84.52 (2)	82.52 (3)	**85.48 (1)**	292	72	131
	wheat	89.54 (3)	89.50 (4)	89.55 (2)	**90.27 (1)**	64	29	48
UCI	Abalone	**100 (1)**	**100 (1)**	**100 (1)**	**100 (1)**	18	4	6
	Breast	98.33 (2)	97.52 (4)	98.33 (2)	**98.84 (1)**	4	1	1
	letter	99.28 (3)	99.42 (2)	99.28 (3)	**99.54 (1)**	83	5	6
	Satimage	**83.57 (1)**	82.61 (4)	82.76 (3)	82.92 (2)	219	18	17

"Batch" corresponds to the classical SVM learning in batch setting without resampling. The numbers in parentheses denote the rank of the corresponding method in the dataset.

of the virtual instances created by SMOTE becomes support vectors. Therefore, SMOTE spends much time to create virtual instances that will not be used in the model. On the other hand, VIRTUAL has already a short training time and uses this time to create more informative virtual instances. In Table 6.4, the numbers in parentheses give the ranks of the g-means prediction performance of the four approaches. The values in bold correspond to a win and VIRTUAL wins in nearly all datasets. The Wilcoxon-signed-rank test (two-tailed) between VIRTUAL and its nearest competitor SMOTE reveals that the zero median hypothesis can be rejected at the significance level 1% ($p = 4.82 \times 10^{-4}$), implying that VIRTUAL performs statistically better than SMOTE in these 14 datasets. These results demonstrate the importance of creating synthetic samples from the informative examples rather than all the examples.

6.5 DIFFICULTIES WITH EXTREME CLASS IMBALANCE

Practical applications rarely provide us with data that have equal numbers of training instances of all the classes. However, in many applications, the imbalance in the distribution of naturally occurring instances is extreme. For example, when labeling web pages to identify specific content of interest, uninteresting pages may outnumber interesting ones by a million to one or worse (consider identifying web pages containing hate speech, in order to keep advertisers off them, cf. [33]).

The previous sections have detailed the techniques that have been developed to cope with moderate class imbalance. However, as class imbalances tends toward the extreme, AL strategies can fail completely—and this failure is not simply due to the challenges of learning models with skewed class distributions, which has received a good bit of study and has been addressed throughout this book. The lack of labeled data compounds the problem because techniques cannot concentrate on the minority instances, as the techniques are unaware which instances to focus on.

Figure 6.10 compares the AUC of logistic regression text classifiers induced by labeled instances selected with uncertainty sampling and with RS. The learning

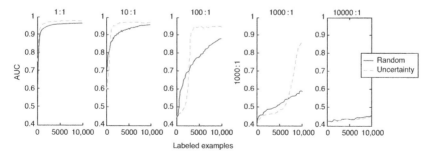

Figure 6.10 Comparison of random sampling and uncertainty sampling on the same dataset with induced skews ranging from 1 : 1 to 10,000 : 1.

task is to differentiate sports web pages from nonsports pages. Depending on the source of the data (e.g., different impression streams from different online advertisers), one could see very different degrees of class skew in the population of relevant web pages. The panels in Figure 6.10, left-to-right, depict increasing amounts of induced class skew. On the far left, we see that for a balanced class distribution, uncertainty sampling is indeed better than RS. For a 10 : 1 distribution, uncertainty sampling has some problems very early on, but soon does better than RS—even more so than in the balanced case. However, as the skew begins to get large, not only does RS start to fail (it finds fewer and fewer minority instances, and its learning suffers), uncertainty sampling does substantially worse than random for a considerable amount labeling expenditure. In the most extreme case shown,[6] both RS and uncertainty sampling simply fail completely. RS effectively does not select any positive examples, and neither does uncertainty sampling.[7]

A practitioner well versed in the AL literature may decide he/she should use a method other than uncertainty sampling in such a highly skewed domain. A variety of techniques have been discussed in Sections 6.2–6.4 for performing AL specifically under class imbalance, including [18–21, 35], as well as for performing density-sensitive AL, where the geometry of the problem space is specifically included when making selections, including [13–15, 17, 36]. While initially appealing, as problems become increasingly difficult, these techniques may not provide results better than more traditional AL techniques—indeed class skews may be sufficiently high to thwart these techniques completely [33].

As discussed later in Section 6.8.1, Attenberg and Provost [33] proposed an alternative way of using human resources to produce labeled training set, specifically tasking people with finding class-specific instances ("guided learning") as opposed to labeling specific instances. In some domains, finding such instances may even be cheaper than labeling (per instance). Guided learning can be much more effective per instance acquired; in one of the Attenberg and Provost's experiments, it outperformed AL as long as searching for class-specific instances was less than eight times more expensive (per instance) than labeling selected instances. The generalization performance of guided learning is shown in Figure 6.12, discussed in Section 6.8.1 for the same setting as Figure 6.10.

6.6 DEALING WITH DISJUNCTIVE CLASSES

Even more subtly still, certain problem spaces may not have such an extreme class skew, but may still be particularly difficult because they possess important but very small disjunctive subconcepts, rather than simple continuously dense

---

[6]10,000 : 1—still orders of magnitude less skewed than some categories.
[7]The curious behavior of AUC< 0.5 here is due to overfitting. Regularizing the logistic regression "fixes" the problem, and the curve hovers about 0.5. See another article in this issue for more insight on models exhibiting AUC< 0.5 [34].

regions of minority and majority instances. Prior research has shown that such "small disjuncts" can comprise a large portion of a target class in some domains [37]. For AL, these small subconcepts act as same as rare classes: if a learner has seen no instances of the subconcept, how can it "know" which instances to label? Note that this is not simply a problem of using the wrong loss function: in an AL setting, the learner does not even know that the instances of the subconcept are misclassified if no instances of a subconcept have yet been labeled. Nonetheless, in a research setting (where we know all the labels), using an undiscriminative loss function, such as classification accuracy or even the AUROC, may result in the researcher not even realizing that an important subconcept has been missed.

To demonstrate how small disjuncts influence (active) model learning, consider the following text classification problem: separating the *Science* articles from the *non-Science* articles within a subset of the 20 newsgroups benchmark set (with an induced class skew of 80−1). Figure 6.11 examines graphically the relative positions of the minority instances through the AL. The black curve shows the AUC (right vertical axis) of the models learned by a logistic regression classifier using uncertainty sampling, rescaled as follows. At each epoch, we sort all instances by their predicted probability of membership in the majority class, $\hat{P}(y = 0|x)$. The black dots in Figure 6.11 represent the minority class instances, with the value on the left vertical axis showing their relative position in this sorted list. The x-axis shows the AL epoch (here each epoch requests 30 new instances from the pool). The black trajectories mostly show instances' relative

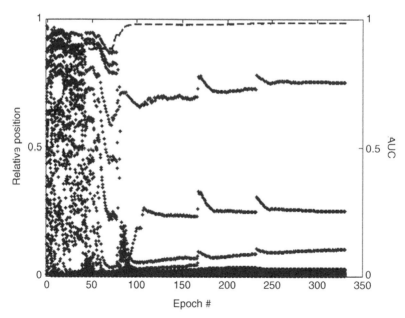

Figure 6.11 A comparison of the learned model's ordering of the instance pool along with the quality of the cross-validated AUC.

positions changing. Minority instances drop down to the very bottom (certain minority) either because they get chosen for labeling, or because labeling some other instance caused the model to "realize" that they are minority instances.

We see that, early on, the minority instances are mixed all throughout the range of estimated probabilities, even as the generalization accuracy increases. Then the model becomes good enough that, abruptly, few minority class instances are misclassified (above $\hat{P} = 0.5$). This is the point where the learning curve levels off for the first time. However, notice that there still are some residual misclassified minority instances, and in particular that there is a cluster of them for which the model is relatively certain they are *majority* instances. It takes many epochs for the AL to select one of these, at which point the generalization performance increases markedly—apparently, this was a subconcept that was strongly misclassified by the model, and so it was not a high priority for exploration by the AL.

On the 20 newsgroups dataset, we can examine the minority instances for which $\hat{P}$ decreases the most in that late rise in the AUC curve (roughly, they *switch* from being misclassified on the lower plateau to being correctly classified afterward). Recall that the minority (positive) class here is "Science" newsgroups. It turns out that these late-switching instances are members of the cryptography (sci.crpyt) subcategory. These pages were classified as non-Science presumably because before having seen any positive instances of the subcategory, they looked much more the same as the many computer-oriented subcategories in the (much more prevalent) non-Science category. As soon as a few were labeled as Science, the model generalized its notion of Science to include this subcategory (apparently pretty well).

Density-sensitive AL techniques did not improve on uncertainty sampling for this particular domain. This was surprising, given the support we have just provided for our intuition that the concepts are disjunctive. One would expect a density-oriented technique to be appropriate for this domain. Unfortunately, in this domain—and we conjecture that this is typical of many domains with extreme class imbalance—the *majority* class is *even more disjunctive* than the minority class. For example, in 20 newsgroups, Science indeed has four very different subclasses. However, non-Science has 16 (with much more variety). Techniques that, for example, try to find as-of-yet unexplored clusters in the instance space are likely to select from the vast and varied majority class. We need more research on dealing with highly disjunctive classes, especially when the less interesting[8] class is more varied than the main class of interest.

6.7 STARTING COLD

The *cold start problem* has long been known to be a key difficulty in building effective classifiers quickly and cheaply via AL [13, 16]. Since the quality of

---

[8]How interesting a class is if it could be measured by its relative misclassification cost, for example.

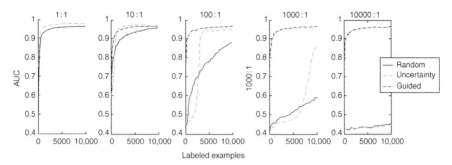

Figure 6.12 Comparison of random sampling and uncertainty sampling and guided learning on the problem shown in Figure 6.10.

data selection directly depends on the understanding of the space provided by the "current" model, early stages of acquisitions can result in a vicious cycle of uninformative selections, leading to poor quality models and therefore to additional poor selections.

The difficulties posed by the cold start problem can be particularly acute in highly skewed or disjunctive problem spaces; informative instances may be difficult for AL to find because of their variety or rarity, potentially leading to substantial waste in data selection. Difficulties early in the AL process can, at least in part, be attributed to the base classifier's poor understanding of the problem space. This cold start problem is particularly acute in otherwise difficult domains. Since the value of subsequent label selections depends on base learner's understanding of the problem space, poor selections in the early phases of AL propagate their harm across the learning curve.

In many research papers, AL experiments are "primed" with a preselected, often class-balanced training set. As pointed out by Attenberg and Provost [33], if the possibility and procedure exist to procure a class-balanced training set to start the process, maybe the most cost-effective model-development alternative is not to do AL at all, but to just continue using this procedure. This is exemplified in Figure 6.12 [33], where the dot-and-hatched lines show the effect of investing resources to continue to procure a class-balanced, but otherwise random, training set (as compared with the active acquisition shown in Figure 6.10).

6.8 ALTERNATIVES TO ACTIVE LEARNING FOR IMBALANCED PROBLEMS

In addition to traditional label acquisition for unlabeled examples, there are other sorts of data that may be acquired at a cost for the purpose of building or improving statistical models. The intent of this section is to provide the reader with a brief overview of some alternative techniques for active data acquisition for predictive model construction in a cost-restrictive setting. We begin this setting with a discussion of class-conditional example acquisition, a paradigm related to AL where examples are drawn from some available unlabeled pool in accordance to some predefined class proportion. We then go on into Section 6.8.2 to touch on active feature labeling (AFL) and active dual supervision (ADS). These two paradigms attempt to replace or supplement traditional supervised learning with class-specific associations on certain feature values. While this set of techniques requires specialized models, significant generalization performance can often be achieved at a reasonable cost by leveraging explicit feature/class relationships. This is often appealing in the active setting, where it is occasionally less challenging to identify class-indicative feature values than it is to find quality training data for labeling, particularly in the imbalanced setting.

6.8.1 Class-Conditional Example Acquisition

Imagine as an alternative to the traditional AL problem setting, where an oracle is queried in order to assign examples to specially selected unlabeled examples, a setting where an oracle is charged with selecting exemplars from the underlying problem space in accordance to some predefined class ratio. Consider as a motivational example, the problem of building predictive models based on data collected through an "artificial nose" with the intent of "sniffing out" explosive or hazardous chemical compounds [38–40]. In this setting, the reactivity of a large number of chemicals is already known, representing label-conditioned pools of available instances. However, producing these chemicals in a laboratory setting and running the resultant compound through the artificial nose may be an expensive, time-consuming process. While this problem may seem quite unique, many data acquisition tasks may be cast into a similar framework.

A much more general issue in selective data acquisition is the amount of control ceded to the "oracle" doing the acquisition. The work discussed so far assumes that an oracle will be queried for some specific value, and the oracle simply returns that value. However, if the oracle is actually a person, he or she may be able to apply considerable intelligence and other resources to "guide" the selection. Such guidance is especially helpful in situations where some aspect of the data is rare—where purely data-driven strategies are particularly challenged.

As discussed throughout this work, in many practical settings, one class is quite rare. As an example motivating the application of class-conditional example acquisition in practice, consider building a predictive model from scratch designed to classify web pages containing a particular topic of interest. While large absolute numbers of such web pages may be present on the web, they may be outnumbered by uninteresting pages by a million to one or worse (take, for instance, the task of detecting and removing hate speech from the web [33]). As discussed in Section 6.5, such extremely imbalanced problem settings present a particularly insidious difficulty for traditional AL techniques. In a setting with a 10,000 : 1 class ratio, a reasonably large labeling budget could be expended without observing a single minority example.[9]

---

[9]Note that in practice, such extremely imbalanced problem settings may actually be quite common.

Previously, we have discussed a plethora of AL techniques specifically tuned for the high skew setting [18–21] as well as techniques where the geometry and feature density of the problem space are explicitly included when making instance selections [13–15, 17, 35, 36]. These techniques, as initially appealing as they may seem, may fail just as badly as traditional AL techniques. Class skew and subconcept rarity discussed in Section 6.6 may be sufficient to thwart them completely [33, 41].

However, in many of these extremely difficult settings, we can task humans to search the problem space for rare cases, using tools (such as search engines) and possibly interacting with the base learner. Consider the motivating example of hate speech classification on the web (from above). While an active learner may experience difficulty in exploring the details of this rare class, a human oracle armed with a search interface is likely to expose examples of hate speech quite easily. In fact, given the coverage of modern web search engines, a human can produce interesting examples from a much larger sample of the problem space far beyond that which is likely to be contained in a sample pool for AL. This is critical due to hardware-imposed constraints on the size of the pool that an active learner is able to choose from—for example, a random draw of several hundred thousand examples from the problem space may not even contain any members of the minority class or of rare disjuncts!

Guided learning is the general process of utilizing oracles to search the problem space, using their domain expertise to *seek* instances representing the interesting regions of the problem space. Figure 6.13 presents the general guided learning setting. Here, given some interface enabling the search over the domain in question, an oracle searches for interesting examples, which are either supplemented with an implicit label by the oracle, or sent for explicit labeling as a second step. These examples are then added to the training set and a model is retrained. Oracles can leverage their background knowledge of the problem being faced. In addition to simply being charged with the acquisition of class-specific examples,

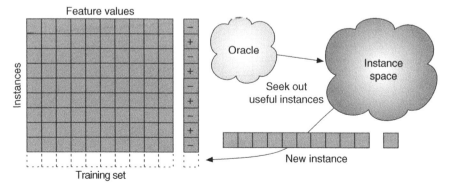

Figure 6.13 Guided learning: an oracle selecting useful examples from the instance space.

by allowing the oracle to interact with the base learner, confusing instances, those that "fool" the model can be sought out from the problem space and used for subsequent training in the form of human-guided uncertainty sampling. This interaction with the base learner can be extended a step further—by allowing the humans to challenge the predictive accuracy of the problem space may potentially reveal "problem areas," portions of the example space where the base model performs poorly that might not be revealed through traditional techniques such as cross-validation studies [42].

Guided learning, along with alternative problem settings such as that faced by the artificial nose discussed earlier deals with situations where an oracle is able to provide "random" examples in arbitrary class proportions. It now becomes interesting to consider just what this class proportion should be? This problem appears to face the inverse of the difficulties faced by AL—labels essentially come for free, while the independent feature values are *completely* unknown and must be gathered at a cost. In this setting, it becomes important to consider the question: "In what proportion should classes be represented in a training set of a certain size?" [43].

Let us call the problem of proportioning class labels in a selection of n additional training instances, "active class selection" (ACS) [38–40, 43]. This process is exemplified in Figure 6.14. In this setting, large, class-conditioned (virtual) pools of available instances with completely hidden feature values are assumed. At each epoch, t, of the ACS process, the task is to leverage the current model when selecting examples from these pools in a proportion believed to have the greatest effectiveness for improving the generalization performance

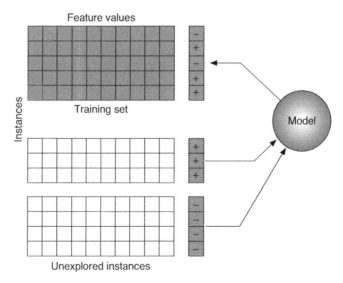

Figure 6.14 Active class selection: gathering instances from random class-conditioned fonts in a proportion believed to offer greatest improvement in generalization performance.

of this model. The feature values for each instance are then collected and the complete instances are added to the training set. The model is reconstructed and the processes is repeated until n examples are obtained (because the budget is exhausted or some other stopping criterion is met, such as a computational limit). Note that this situation can be considered to be a special case of the instance completion setting of active feature-value acquisition (cf. [44]). It is a degenerate special case because, before the selection, there is no information at all about the instances other than their classes.

For the specific problem at the heart of ACS, the extreme lack of information to guide selection leads to the development of unique uncertainty and utility estimators, which, in the absence of predictive covariates, require unique approximations.[10] While alternative approaches to ACS have emerged, for thematic clarity, uncertainty-based and expected-utility-based approaches will be presented first. Note that because effective classification requires that both sides of a prediction boundary be represented, unlike typical AL techniques, ACS typically *samples* classes from their respective score distributions [45, 46].

6.8.1.1 Uncertainty-Based Approaches
This family of techniques for performing ACS is based on the volatility in the predictions made about certain classes—those classes whose cross-validated predictions are subject to the most change between successive epochs of instance selection are likely to be based on an uncertain predictor and amenable to refinement by the incorporation of additional training data [38, 40]. Analogous to the case of more traditional uncertainty-based data acquisition, several heuristics have been devised to capture the notion of variability.

One measure of the uncertainty of a learned model is how volatile its predictive performance is in the face of new training data. Take a typical learning curve, for instance, those presented in Figure 6.6. Notice that the modeling is much more volatile at the left side of the figure, showing large changes in generalization performance for the same amount of new training data. We can think that as the predictor gains knowledge of the problem space, it tends to solidify in the face of data, exhibiting less change and greater certainty For ACS, we might wonder if the learning curves will be equally steep regardless of the class of the training data [38–40]. With this in mind, we can select instances at epoch t from the classes in proportion to their improvements in accuracy at $t - 1$ and $t - 2$. For example, we could use cross-validation to estimate the generalization performance of the classifier with respect to each class, $\mathcal{A}(c)$; class c can then be sampled according to:

$$p_{\mathcal{A}}^{t}(c) \propto \frac{\max\left\{0, \mathcal{A}^{t-1}(c) - \mathcal{A}^{t-2}(c)\right\}}{\sum_{c'} \max\left\{0, \mathcal{A}^{t-1}(c') - \mathcal{A}^{t-2}(c')\right\}},$$

[10] In realistic settings, for instance, such as potential application for ACS, guided learning, this lack of information assumption may be softened.

Alternatively, we could consider general volatility in class members' predicted labels, beyond improvement in the model's ability to predict the class. Again, using cross-validated predictions at successive epochs, it is possible to isolate members of each class, and observe changes in the predicted class for each instance. For example, when the predicted label of a given instance changes between successive epochs, we can deem the instance to have been *redistricted* [38–40]. Again considering the level of volatility in a model's predictions to be a measurement of uncertainty, we can sample classes at epoch t according to each classes' proportional measure of redistricting:

$$p_{\mathcal{R}}^t(c) \propto \frac{\frac{1}{|c|} \sum_{x \in c} \mathbb{I}(f^{t-1}(x) \neq f^{t-2}(x))}{\sum_{c'} \frac{1}{|c'|} \sum_{x \in c'} \mathbb{I}(f^{t-1}(x) \neq f^{t-2}(x))},$$

where $\mathbb{I}(\cdot)$ is an indicator function taking the value of 1 if its argument is true and 0 otherwise. $f^{t-1}(x)$ and $f^{t-2}(x)$ are the predicted labels, for instance, x from the models trained at epoch $t-1$ and $t-2$, respectively [38–40].

6.8.1.2 Expected Class Utility The previously described ACS heuristics are reliant on the assumption that adding examples belonging to a particular class will improve the predictive accuracy with respect to that class. This does not directly estimate the utility of adding members of a particular class to a model's overall performance. Instead, it may be preferable to select classes whose instances' presence in the training set will reduce a model's misclassification cost by the greatest amount in expectation.

Let $\text{cost}(c_i|c_j)$ be the cost of predicting c_i on an instance x whose true label is c_j. Then the expected empirical misclassification cost over a sample dataset, $\mathbb{D}$, is:

$$\hat{R} = \frac{1}{|\mathbb{D}|} \sum_{x \in \mathbb{D}} \sum_i \hat{P}(c_i|x)\text{cost}(c_i|y),$$

where y is the correct class for a given x. Typically in the ACS setting, this expectation would be taken over the training set (e.g., $\mathbb{D} = T$), preferably using cross-validation. In order to reduce this risk, we would like to select examples from class c, leading to the greatest reduction in this expected risk [39].

Consider a predictive model $\hat{P}^{T \cup c}(\cdot|x)$, a model built on the training set, T, supplemented with an arbitrary example belonging to class c. Given the opportunity to choose an additional class-representative example to the training pool, we would like to select the class that reduces the expected risk by the greatest amount:

$$\bar{c} = \arg\max_c U(c),$$

where

$$U(c) = \frac{1}{|\mathbb{D}|} \sum_{x \in \mathbb{D}} \sum_i \hat{P}^T(c_i|x)\text{cost}(c_i|y) - \frac{1}{|\mathbb{D}|} \sum_{x \in \mathbb{D}} \sum_i \hat{P}^{T \cup c}(c_i|x)\text{cost}(c_i|y).$$

Of course the benefit of adding additional examples on a test dataset is unknown. Furthermore, the impact of a particular class's examples may vary depending on the feature values of particular instances. In order to cope with these issues, we can estimate via cross-validation on the training set. Using sampling, we can try various class-conditional additions and compute the expected benefit of a class across that class's representatives in T, assessed on the testing folds. The earlier-mentioned utility then becomes:

$$\hat{U}(c) = E_{x \in c}\left[\frac{1}{|\mathbb{D}|}\sum_{x \in \mathbb{D}}\sum_{i} \hat{P}^T(c_i|x)\text{cost}(c_i|y) - \frac{1}{|\mathbb{D}|}\sum_{x \in \mathbb{D}}\sum_{i} \hat{P}^{T \cup c}(c_i|x)\text{cost}(c_i|y)\right].$$

Note that it is often preferred to add examples in batch. In this case, we may wish to sample from the classes in proportion to their respective utilities:

$$p_{\hat{U}}^t(c) \propto \frac{\hat{U}(c)}{\sum'_c \hat{U}(c)'}.$$

Further, diverse class-conditional acquisition costs can be incorporated, utilizing $\hat{U}(c)/\omega_c$ in place of $\hat{U}(c)$, where ω_c is the (expected) cost of acquiring the feature vector of an example in class c.

6.8.1.3 Alternative Approaches to ACS

In addition to uncertainty-based and utility-based techniques, there are several alternative techniques for performing ACS. Motivated by empirical results showing that barring any domain-specific information, when collecting examples for a training set of size n, a balanced class distribution tends to offer reasonable AUC on test data [43, 47], a reasonable baseline approach to ACS is simply to select classes in balanced proportion.

Search strategies may alternately be employed in order to reveal the most effective class ratio at each epoch. Utilizing a nested cross-validation on the training set, the space of class ratios can be explored, with the most favorable ratio being utilized at each epoch. Note that it is not possible to explore all possible class ratios in all epochs, without eventually spending too much on one class or another. Thus, as we approach n, we can narrow the range of class ratios, assuming that there is a problem-optimal class ratio that will become more apparent as we obtain more data [43].

It should be noted that many techniques employed for building classification models assume an identical or similar training and test distribution. Violating this assumption may lead to biased predictions on test data where classes preferentially represented in the training data are predicted more frequently. In particular, increasing the prior probability of a class increases the posterior probability of the class, moving the classification boundary for that class so that more cases are classified into that class" [48, 49]. Thus in settings where instances are selected specifically in proportions different from those seen in the wild, posterior probability estimates should be properly calibrated to be aligned with the test data, if possible [43, 49, 50].

Prior work has repeatedly demonstrated the benefits of performing ACS beyond simply selecting random examples from an example pool for acquisition or simply using uniformly balanced selection. However, in many cases, simply casting what would typically be an AL problem into an ACS problem, and selecting examples uniformly among the classes can provide results far better than what would be possible with AL alone. For instance, the learning curves presented in Figure 6.12 compare such uniform guided learning with AL and simple random selection. Providing the model with an essentially random but class-balanced training set far exceeds the generalization performance possible for an AL strategy or by random selection once the class skew becomes substantial. More intelligent ACS strategies may make this difference even more pronounced, and should be considered if the development effort associated with incorporating such strategies would be outweighed by the savings coming from reduced data acquisition costs.

6.8.2   Feature-Based Learning and Active Dual Supervision

While traditional supervised learning is by far the most prevalent classification paradigm encountered in the research literature, it is not the only approach for incorporating human knowledge into a predictive system. By leveraging, for instance, class associations with certain feature values, predictive systems can be trained that offer potentially excellent generalization performance without requiring the assignment of class labels to individual instances. Consider the example domain of predicting the sentiment of movie reviews. In this context, it is clear that the presence of words such as "amazing" and "thrilling" carries an association with the positive class, while terms such as "boring" and "disappointing" evoke negative sentiment [51]. Gathering this kind of annotation leverages an oracle's prior experience with the class polarity of certain feature values—in this case, the emotion that certain terms tend to evoke. The systematic selection of feature values for labeling by a machine learning system is referred to as *active feature-value labeling*,[11]. The general setting where class associations are actively sought for both feature values and particular examples is known as *ADSs*. The process of selection for AFL and ADS is shown in Figures 6.15 and 6.16, respectively.

Of course, incorporating the class polarities associated with certain feature values typically requires specialized models whose functional form has been designed to leverage feature-based background knowledge. While a survey of models for incorporating such feature- value/class polarities is beyond the scope of this chapter, an interested reader is advised to seek any number of related papers (cf. [52–58]). However, while sophisticated models of this type have

---

[11] For brevity, this is often shortened as AFL, a moniker that is best suited for domains with binary features.

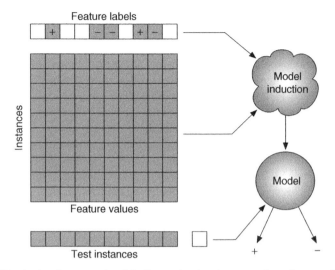

Figure 6.15 Active feature-value labeling: selecting feature values for association with a certain class polarity.

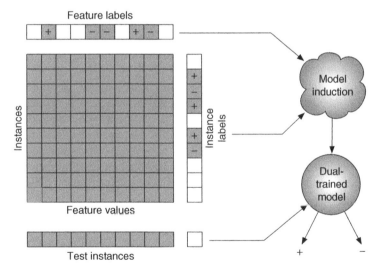

Figure 6.16 Active dual supervision: simultaneous acquisition of label information for both feature values and instances.

been devised, practical, yet very simple solutions are easy to imagine. Consider at the most basic extreme class assignment rules, where the presence of certain feature values (or combinations of feature values) connects an example with a particular class. A rudimentary "dual" model can also be constructed by simply averaging the class predictions of a traditional supervised machine learning model with a model based on feature-value/class relationships.

Given the option of incorporating prior knowledge associated with certain feature values into the predictive behavior of a model, the question then becomes: *which feature examples should be selected for labeling?* Initially, this may seem to be a deflection, replacing the search for useful information of one kind with another. However, there are several reasons why seeking feature values to "label" may be preferred to labeling examples. The most obvious reason for selecting feature values for labeling is that traditional is often "slow." It may take many labeled movie reviews to teach the model that "amazing" has a positive association, and not the other uninformative terms that just happen to occur in positive reviews by coincidence. In this way, the cold start problem faced in AL[12] may be less acute in AFL and ADS—while complex feature/class relationships may be difficult to achieve using AFL; reasonable generalization performance is often achievable with few requests to an oracle. Second, the labor costs associated with assigning a class polarity to a certain feature value is often quite low—it is easy for a human to associate the term *terrible* with negative movie reviews, while labeling one particular movie review as positive or negative requires reading the entire document. Note that of course not every term is polar; in the ongoing example of movie reviews, it is easy to imagine that a few terms have a natural association with the positive or negative class, while most terms on their own do not have such a polarity. However, this imbalance between polar and non-polar feature values is often far less acute than the imbalance between classes in many machine learning problems. A problem domain with a class imbalance of 1,000,000 : 1 may still have 1 in 10 features exhibiting some meaningful class association. Perhaps even more importantly, the ratio between positively and negatively linked feature values (for instance) may be far more balanced than the ratio between those classes in the wild. In fact, there is not necessarily a relationship between the base rate and the ratio of strongly identifiable positively and negatively associated feature values. While selecting useful feature values in practice is still often a challenging problem, experience has shown that it is often more informative to select random features for labeling than random examples. More intelligent selection heuristics can make this preference for AFL and ADS even stronger.

Again, giving a thorough survey of selection heuristics for performing AFL is beyond the scope of this chapter. However, we will provide the reader with a brief overview of the techniques typically employed for such tasks. As in many selective data acquisition tasks for machine learning, we see two common themes: uncertainty-based selection and expected-utility-based approaches. In the following, we will briefly present some more popular techniques for AFL delineated accordingly. We will then briefly discuss the techniques for ADS, selecting features and examples for labeling simultaneously.

6.8.2.1 Uncertainty-Based AFL   By far the most prevalent class of AL heuristics, uncertainty-based approaches do not have a direct analogue to the selection

---

[12]Recall Section 6.7

problem faced in ADS; feature-value/class associations factor quite differently into the formation and training of machine learning models than more traditional example labels. Nonetheless, approaches thematically similar to the ADS problem have been developed, where features are selected to labeling according to some notion of stability within the current model. As in traditional uncertainty selection in AL, the primary difference among these uncertainty-based ADS techniques is the manner by which feature uncertainty is estimated. In the simplest case, using a linear model, the coefficients on particular feature values may be interpreted as a measure of uncertainty, with lower coefficient magnitude corresponding to a greater degree of uncertainty [59]. The Naíve Bayes-like model used presented in [52] presents a more appealing option—feature-value label uncertainty can be measured by looking at the magnitude of the log-odds of the feature-value likelihoods: $|\log p(f|+)/p(f|-)|$ for feature value f and classes $+$ and $-$. Again, a smaller value corresponds to increased uncertainty.

A range of other techniques for uncertainty-based AFL exist. By the creation of one-term pseudo-documents, Godbole et al [60] coerce the notion of feature label uncertainty into a more traditional instance uncertainty framework for text classification tasks. By incorporating the label information on each feature value with unlabeled examples, Druck et al. [58] create a corresponding *generalized expectation* term that rates the model's predicted class distribution conditioned on the presence of the particular feature. This rating penalizes these predicted class distributions according to their KL-divergence from reference distributions constructed using labeled features. Similarly, Liang et al. [61] learn from labeled examples and actively selected constraints in the form of expectations with some associated noise from particular examples. Druck et al. [62] analyze several uncertainty-based selection techniques for gathering feature labels when training conditional random fields, finding that the total uncertainty (measured as the sum of the marginal entropies) tends to favor more frequent features. As a remedy, they propose an uncertainty scheme where the mean uncertainty is weighted by the log of the counts of the associated feature values.

Interestingly, even for reasonable uncertainty estimators, feature/class uncertainty may not be a desirable criterion for selection. Consider the discussion made previously regarding the preponderance of uninformative features. Clearly, in the case of document classification, terms such as "of" and "the" will seldom have any class polarity. At the same time, these terms are likely to have a high degree of uncertainty, leaving uncertainty-based approaches to perform poorly in practice. Preferring to select features based on *certainty*, that is, selecting those features with the least uncertainty, seems to work much better in practice [59, 63].

6.8.2.2 *Expected Utility-Based AFL*

As with traditional uncertainty (and certainty) sampling for AL, the query corresponding to the greatest level of uncertainty may not necessarily be the query offering the greatest level of information to the model. This is particularly true in noisy or complex environments. Instead of using a heuristic to estimate the information value of a particular feature label, it is possible to estimate this quantity directly. Let q enumerate over all possible feature values that may be queried for labels. We can estimate the expected utility of such a query by: $EU(q) = \sum_{k=1}^{K} P(q = c_k)\mathcal{U}(q = c_k)/\omega_q$, where $P(q = c_k)$ is the probability of the instance or feature queried being associated with class c_k, ω_q is the cost of query q, and $\mathcal{U}$ is some measure of the utility of q.[13] This results in the decision-theoretic optimal policy, which is to ask for feature labels which, once incorporated into the data, will result in the highest increase in classification performance in *expectation* [51, 63].

6.8.2.3 Active Dual Supervision   ADS is concerned with situations where it is possible to query an oracle for labels associated with both feature values and examples. Even though such a paradigm is concerned with the simultaneous acquisition of feature and example labels, the simplest approach is treating each acquisition problem separately and then mixing the selections somehow. Active interleaving performs a separate (un) certainty-based ordering on features and on examples, and chooses selections from the top of each ordering according to some predefined proportion. The different nature of feature value and example uncertainty values lead to incompatible quantities existing on different scales, preventing a single, unified ordering. However, expected utility can be used to compute a single unified metric, encapsulating the value of both types of data acquisition. As mentioned earlier, we are estimating the utility of a certain feature of example query q as: $EU(q) = \sum_{k=1}^{K} P(q = c_k)\mathcal{U}(q = c_k)/\omega_q$. Using a single utility function for both features and examples and incorporating label acquisition costs, costs and benefits of the different types of acquisition can be optimized directly [51].

6.9   CONCLUSION

This chapter presents a broad perspective on the relationship between AL—the selective acquisition of labeled examples for training statistical models—and imbalanced data classification tasks where at least one of the classes in the training set is represented with much fewer instances than the other classes. Our comprehensive analysis of this relationship leads to the identification of two common associations, namely (i) is the ability of AL to deal with the data imbalance problem that, when manifested in a training set, typically degrades the generalization performance of an induced model, and (ii) is the impact class imbalance may have on the abilities of an otherwise reasonable AL scheme to select informative examples, a phenomenon that is particularly acute as the imbalance tends toward the extreme.

Mitigating the impact of class imbalance on the generalization performance of a predictive model, in Sections 6.3 and 6.4, we present AL as an alternative to more conventional resampling strategies. An AL strategy may select a dataset

---

[13] For instance, cross-validated accuracy or log-gain may be used.

that is both balanced and extremely informative in terms of model training. Early stopping criteria for halting the selection process of AL can further improve the generalization of induced models. That is, a model trained on a small but informative subsample often offers performance far exceeding what can be achieved by training on a large dataset drawn from the natural, skewed base rate. The abilities of AL to provide small, balanced training from large, but imbalanced problems are enhanced further by VIRTUAL, introduced in Section 6.4. Here, artificially generated instances supplement the pool of examples available to the active selection mechanism.

Additionally, it is noted throughout that abilities of an AL system tend to degrade as the imbalance of the underlying distribution increases. While at more moderate imbalances, the quality of the resulting training set may still be sufficient to provide usable statistical models, but at more substantial class imbalances, it may be difficult for a system based on AL to produce accurate models. Throughout Sections 6,2–6.4, we illustrate a variety of AL techniques specially adapted for imbalanced settings, techniques that may be considered by practitioners facing difficult problems. In Sections 6.5 and 6.6, we note that as a problem's class imbalance tends toward the extreme, the selective abilities of an AL heuristic may fail completely. We present several alternative approaches for data acquisition in Section 6.8, mechanisms that may alleviate the difficulties AL faces in problematic domains. Among these alternatives are guided learning and ACS in Section 6.8.1, and using associations between specific feature values and certain classes in Section 6.8.2.

Class imbalance presents a challenge to statistical models and machine learning systems in general. Because the abilities of these models are so tightly coupled with the data used for training, it is crucial to consider the selection process that generates this data. This chapter discusses specifically this problem. It is clear that when building models for challenging imbalanced domains, AL is an aspect of the approach that should not be ignored.

REFERENCES

1. B. Settles, "Active learning literature survey," Computer Sciences Technical Report 1648, University of Wisconsin-Madison, 2009.

2. V. V. Federov, *Theory of Optimal Experiments*. New York: Academic Press, 1972.

3. D. Cohn, L. Atlas, and R. Ladner, "Improving generalization with active learning," *Machine Learning*, vol. 15, pp. 201–221, 1994.

4. D. D. Lewis and W. A. Gale, "A sequential algorithm for training text classifiers," in *SIGIR '94: Proceedings of the 17th Annual International ACM SIGIR Conference on Research and Development in Information Retrieval* (New York, NY, USA), pp. 3–12, Springer-Verlag Inc., 1994.

5. Y. Freund, "Sifting informative examples from a random source," in *Working Notes of the Workshop on Relevance, AAAI Fall Symposium Series*, pp. 85–89, 1994.

6. I. Dagan and S. P. Engelson, "Committee-based sampling for training probabilistic classifiers," in *Proceedings of the Twelfth International Conference on Machine Learning* (Tahoe City, CA, USA), pp. 150–157, Morgan Kaufmann, 1995.

7. Y. Freund, H. S. Seung, E. Shamir, and N. Tishby, "Information, prediction, and query by committee," in *Advances in Neural Information Processing Systems 5, [NIPS Conference]*, pp. 483–490, Morgan Kaufmann, 1993.

8. S. Tong and D. Koller, "Support vector machine active learning with applications to text classification," *Journal of Machine Learning Research*, vol. 2, pp. 45–66, 2002.

9. N. Roy and A. McCallum, "Toward optimal active learning through sampling estimation of error reduction," in *ICML*, Morgan Kaufmann, 2001.

10. R. Moskovitch, N. Nissim, D. Stopel, C. Feher, R. Englert, and Y. Elovici, "Improving the detection of unknown computer worms activity using active learning," in *Proceedings of the 30th Annual German Conference on Advances in Artificial Intelligence*, KI '07, pp. 489–493, Springer-Verlag, 2007.

11. Y. Guo and R. Greiner, "Optimistic active learning using mutual information," in *Proceedings of the 20th International Joint Conference on Artificial Intelligence, IJCAI'07* (Hyderabad, India), pp. 823–829, 2007.

12. B. Settles and M. Craven, "An analysis of active learning strategies for sequence labeling tasks," in *Proceedings of the Conference on Empirical Methods in Natural Language Processing*, EMNLP '08 (Jeju Island, Korea), pp. 1070–1079, Association for Computational Linguistics, 2008.

13. J. Zhu, H. Wang, T. Yao, and B. K. Tsou, "Active learning with sampling by uncertainty and density for word sense disambiguation and text classification," in *COLING '08*, Association for Computational Linguistics, 2008.

14. H. T. Nguyen and A. Smeulders, "Active learning using pre-clustering," in *ICML*, ACM, 2004.

15. A. K. Mccallum and K. Nigam, "Employing EM in pool-based active learning for text classification," in *ICML*, 1998.

16. P. Donmez, J. G. Carbonell, and P. N. Bennett, "Dual strategy active learning," in *ECML '07*, Springer, 2007.

17. P. Donmez and J. Carbonell, "Paired sampling in density-sensitive active learning," in *Proceedings of 10th International Symposium on Artificial Intelligence and Mathematics* (Ft. Lauderdale, FL, USA), 2008.

18. K. Tomanek and U. Hahn, "Reducing class imbalance during active learning for named entity annotation," in *K-CAP '09*, 105–112, ACM, 2009.

19. S. Ertekin, J. Huang, L. Bottou, and C. L. Giles, "Learning on the border: Active learning in imbalanced data classification," in *Proceedings of the 16th ACM Conference on Information and Knowledge Management (CIKM)* (Lisbon, Portugal), pp. 127–136, ACM, 2007.

20. M. Bloodgood and K. V. Shanker, "Taking into account the differences between actively and passively acquired data: The case of active learning with support vector machines for imbalanced datasets," in *NAACL '09*, Association for Computational Linguistics, 2009.

21. S. Ertekin, *Learning in Extreme Conditions: Online and Active Learning with Massive, Imbalanced and Noisy Data*. PhD thesis, The Pennsylvania State University, 2009.

22. J. Zhu and E. Hovy, "Active learning for word sense disambiguation with methods for addressing the class imbalance problem," in *EMNLP-CoNLL 2007* (Prague, Czech Republic), 2007.

23. G. Schohn and D. Cohn, "Less is more: Active learning with support vector machines," in *Proceedings of the 17th International Conference on Machine Learning (ICML)* (Stanford, CA, USA), pp. 839–846, Morgan Kaufmann, 2000.

24. A. Bordes, S. Ertekin, J. Weston, and L. Bottou, "Fast kernel classifiers with online and active learning," *Journal of Machine Learning Research*, vol. 6, pp. 1579–1619, 2005.

25. J. Huang, S. Ertekin, and C. L. Giles, "Efficient name disambiguation for large scale datasets," in *Proceedings of European Conference on Principles and Practice of Knowledge Discovery in Databases (ECML/PKDD)* (Berlin, Germany), vol. 4213/2006, pp. 536–544, 2006.

26. V. Vapnik, *The Nature of Statistical Learning Theory*. New York: Springer, 1995.

27. A. J. Smola and B. Schölkopf, "Sparse greedy matrix approximation for machine learning," in *Proceedings of 17th International Conference on Machine Learning (ICML)* (Stanford, CA), pp. 911–918, Morgan Kaufmann, 2000.

28. M. Kubat and S. Matwin, "Addressing the curse of imbalanced training datasets: One sided selection," in *Proceedings of 14th International Conference on Machine Learning (ICML)*, vol. 30, no. 2–3, pp. 195–215, 1997.

29. N. Japkowicz and S. Stephen, "The class imbalance problem: A systematic study," *Intelligent Data Analysis*, vol. 6, no. 5, pp. 429–449, 2002.

30. N. V. Chawla, K. W. Bowyer., L. O. Hall, and W. P. Kegelmeyer, "SMOTE: Synthetic minority over-sampling technique," *Journal of Artificial Intelligence Research*, vol. 16, pp. 321–357, 2002.

31. N. Japkowicz, "The class imbalance problem: Significance and strategies," in *Proceedings of 2000 International Conference on Artificial Intelligence (IC-AI'2000)*, vol. 1, pp. 111–117, 2000.

32. C. X. Ling and C. Li, "Data mining for direct marketing: Problems and solutions," in *Knowledge Discovery and Data Mining*, pp. 73–79, 1998.

33. J. Attenberg and F. Provost, "Why label when you can search? Strategies for applying human resources to build classification models under extreme class imbalance," in *KDD*, pp. 423–432, ACM, 2010.

34. C. Perlich and G. Swirszcz, "On cross-validation and stacking: Building seemingly predictive models on random data," *SIGKDD Explorations*, vol. 12, no. 2, p. 11–15, 2010.

35. J. He and J. G. Carbonell, "Nearest-neighbor-based active learning for rare category detection," in *NIPS* (Vancouver, Canada), pp. 633–640, MIT Press, 2007.

36. Z. Xu, K. Yu, V. Tresp, X. Xu, and J. Wang, "Representative sampling for text classification using support vector machines," in *ECIR*, pp. 393–407, Springer-Verlag, 2003.

37. Weiss, G. M., "The impact of small disjuncts on classifier learning," *Annals of Information Systems*, vol. 8, pp. 193–226, 2010.

38. R. Lomasky, C. Brodley, M. Aernecke, D. Walt, and M. Friedl, "Active class selection," *Machine Learning: ECML*, vol. 4701, pp. 640–647, 2007.

39. R. Lomasky, *Active Acquisition of Informative Training Data*. PhD thesis, Tufts University, 2010.

40. R. Lomasky, C. E. Brodley, S. Bencic, M. Aernecke, and D. Walt, "Guiding class selection for an artificial nose," in *NIPS Workshop on Testing of Deployable Learning and Decision Systems*, 2006.

41. J. Attenberg and F. Provost, "Inactive learning? Difficulties employing active learning in practice," *SIGKDD Explorations*, vol. 12, no. 2, pp. 36–41, 2010.

42. J. Attenberg, P. Ipeirotis, and F. Provost, "Beat the machine: Challenging workers to find the unknown unknowns," in *Proceedings of the 3rd Human Computation Workshop (HCOMP 2011)*, 2011.

43. G. M. Weiss and F. Provost, "Learning when training data are costly: The effect of class distribution on tree induction," *Journal of Artificial Intelligence Research*, vol. 19, pp. 315–354, 2003.

44. P. Melville, F. J. Provost, and R. J. Mooney, "An expected utility approach to active feature-value acquisition," in *International Conference on Data Mining* (Houston, TX, USA), pp. 745–748, 2005.

45. M. Saar-tsechansky and F. Provost, "Active learning for class probability estimation and ranking," in *Proceedings of the Seventeenth International Joint Conference on Artificial Intelligence (IJCAI-2001)*, pp. 911–920, Morgan Kaufmann, 2001.

46. M. Saar-Tsechansky and F. Provost, "Active sampling for class probability estimation and ranking," *Machine Learning*, vol. 54, no. 2, pp. 153–178, 2004.

47. G. Weiss and F. Provost, "The effect of class distribution on classifier learning," 2001.

48. SAS Institute Inc., *Getting Started with SAS Enterprise Miner*. Cary, NC: SAS Institute Inc., 2001.

49. F. Provost, "Machine learning from imbalanced data sets 101," in *Proceedings of the AAAI Workshop on Imbalanced Data Sets*, 2000.

50. C. Elkan, "The foundations of cost-sensitive learning," in *Proceedings of the Seventeenth International Joint Conference on Artificial Intelligence*, pp. 973–978, Morgan Kaufmann, 2001.

51. J. Attenberg, P. Melville, and F. J. Provost, "A unified approach to active dual supervision for labeling features and examples," in *European Conference on Machine Learning*, pp. 40–55, Springer-Verlag, 2010.

52. P. Melville, W. Gryc, and R. Lawrence, "Sentiment analysis of blogs by combining lexical knowledge with text classification," in *KDD*, pp. 1275–1284 ACM, 2009.

53. R. E. Schapire, M. Rochery, M. G. Rahim, and N. Gupta, "Incorporating prior knowledge into boosting," in *International Conference on Machine Learning (ICML)*, Morgan Kaufmann, 2002.

54. X. Wu and R. Srihari, "Incorporating prior knowledge with weighted margin support vector machines," in *Conference on Knowledge Discovery and Data Mining (KDD)*, pp. 326–333, ACM, 2004.

55. B. Liu, X. Li, W. S. Lee, and P. Yu, "Text classification by labeling words," in *AAAI*, pp. 425–430, AAAI Press 2004.

56. A. Dayanik, D. Lewis, D. Madigan, V. Menkov, and A. Genkin, "Constructing informative prior distributions from domain knowledge in text classification," in *SIGIR*, pp. 493–500, ACM, 2006.

57. G. Kunapuli, K. P. Bennett, A. Shabbeer, R. Maclin, and J. Shavlik, "Online knowledge-based support vector machines," in *ECML/PKDD (2)* (J. L. Balczar, F. Bonchi, A. Gionis, and M. Sebag, eds.), vol. 6322, 2010.

58. G. Druck, G. Mann, and A. McCallum, "Learning from labeled features using generalized expectation criteria," in *Special Interest Group in Information Retrieval (SIGIR)*, 595–602, ACM, 2008.

59. V. Sindhwani, P. Melville, and R. Lawrence, "Uncertainty sampling and transductive experimental design for active dual supervision," in *Proceedings of the 26th International Conference on Machine Learning (ICML-09)*, pp. 953–960, ACM, 2009.

60. S. Godbole, A. Harpale, S. Sarawagi, and S. Chakrabarti, "Document classification through interactive supervision of document and term labels," in *Practice of Knowledge Discovery in Databases (PKDD)*, pp. 185–196, Springer-Verlag, 2004.

61. P. Liang, M. I. Jordan, and D. Klein, "Learning from measurements in exponential families," in *ICML*, pp. 641–648, ACM, 2009.

62. G. Druck, B. Settles, and A. McCallum, "Active learning by labeling features," in *Conference on Empirical Methods in Natural Language Processing (EMNLP '09)* (Singapore), pp. 81–90, Association for Computational Linguistics, 2009.

63. P. Melville and V. Sindhwani, "Active dual supervision: Reducing the cost of annotating examples and features," in *Proceedings of the NAACL HLT 2009 Workshop on Active Learning for Natural Language Processing*, Association of Computational Linguistics, 2009.

7

NONSTATIONARY STREAM DATA LEARNING WITH IMBALANCED CLASS DISTRIBUTION

Sheng Chen

Merrill Lynch, Bank of America, New York, NY, USA

Haibo He

Department of Electrical, Computer, and Biomedical Engineering, University of Rhode Island, Kingston, RI, USA

Abstract: The ubiquitous imbalanced class distribution occurring in real-world datasets has stirred considerable interest in the study of *imbalanced learning*. However, it is still a relatively uncharted area when it is a *nonstationary data stream* with imbalanced class distribution that needs to be processed. Difficulties in this case are generally twofold. First, a dynamically structured learning framework is required to catch up with the evolution of unstable class concepts, that is, concept drifts. Second, an imbalanced class distribution over data streams demands a mechanism to intensify the underrepresented class concepts for improved overall performance. For instance, in order to design an intelligent spam filtering system, one needs to make a system that can self-tune its learning parameters to keep pace with the rapid evolution of spam mail patterns and tackle the fundamental problem of normal emails being severely outnumbered by spam emails in some situations; yet it is so much more expensive to misclassify a normal email as spam, for example, confirmation of a business contract, than the other way around. This chapter introduces learning algorithms that were specifically proposed to tackle the problem of learning from nonstationary datasets with imbalanced class distribution. System-level principles and a framework of these methods are described at an algorithmic level, the soundness of which is further validated through theoretical analysis as well as simulations on both synthetic and real-world benchmarks with varied levels of imbalanced ratio and noise.

Imbalanced Learning: Foundations, Algorithms, and Applications, First Edition.
Edited by Haibo He and Yunqian Ma.
© 2013 The Institute of Electrical and Electronics Engineers, Inc. Published 2013 by John Wiley & Sons, Inc.

7.1 INTRODUCTION

Learning from data streams has been featured in many practical applications such as network traffic monitoring and credit fraud identification [1]. Generally speaking, a data stream is a sequence of unbounded, real-time data items with a very high rate that can be read only once by an application [2]. The restriction at the end of this definition is also called a *one-pass constraint* [3], which is also confirmed by other studies [4–6]. The study of learning from data streams has been a quite popular topic in the machine learning community. Domingos and Hulten [7] proposed the very fast decision tree (VFDT) to address data mining from high-speed data streams such as web access data. By using Hoeffding bounds, it can offer an approximately identical performance as a conventional learner on a static dataset. Learn++ [8] approaches learning from data streams through an aggressive ensemble-of-ensemble learning paradigm. Briefly speaking, Learn++ processes the data streams in units of data chunks. For each data chunk, Learn++ applies the base learner of multilayer perceptron (MLP) to create multiple ensemble hypotheses on it. He and Chen [9] proposed the incremental multiple-object recognition and localization (IMORL) framework to address learning from video data streams. It calculates the Euclidean distance in feature space between examples within consecutive data chunks to transmit sampling weights in a biased manner, that is, hard-to-learn examples would gradually be assigned higher weights for learning, which resembles AdaBoost's weight-updating mechanism [10] to some degree. In Literature [11], an approach to real-time generation of fuzzy rule-based systems of eXtended Takagi–Sugeno (xTS) type from data streams was proposed, which applies an incremental clustering procedure to generate clusters to form fuzzy rule-based systems. Georgieva and Filev [12] proposed the Gustafson–Kessel algorithm for incremental clustering of data streams. It applies the adaptive-distance metric to identify clusters with different shapes and orientations. As a follow-up, Filev and Georgieva [13] extended the Gustafson–Kessel algorithm to enable real-time clustering of data streams.

Inability to store all data into memory for learning as done by traditional approaches has not been the sole challenge presented by data streams. As the term concept drift suggests, it is undesirable, yet inevitable, that class concepts evolve as data streams move forward. This property, combined with a virtually unbounded volume of data streams, accounts for the so-called stability–plasticity dilemma [14]. One may be trapped in an endless loop of pondering, either reserving just the most recent knowledge to battle against concept drift or keeping track of knowledge as much as possible in avoidance of "catastrophic forgetting" [14]. Many studies have been recorded to strike a balance on this dilemma. Marked by an effort of adapting an ensemble approach to time-evolving data streams, streaming ensemble algorithm (SEA) [15] maintains an ensemble pool of C4.5 hypotheses with a fixed size, each of which is built on a data chunk with a unique timestamp. When a request to insert a new hypothesis is made but the ensemble pool is fully occupied, some criterion is introduced to evaluate whether the new hypothesis is qualified enough to be accommodated at the expense of removing an existing hypothesis. Directly targeted at making one's choice between the new and old data, Fan [16] examined the necessity of referring to the old data. If it is unnecessary, reserving the most recent data would suffice to yield a hypothesis with satisfactory performance. Otherwise, cross-validation will be applied to locate the portion of old data that is most helpful in complementing the most recent data for building an optimal hypothesis. The potential problem in this approach is the choice of granularity for cross-validation. Finer granularity would more accurately provide the desirable portion of the old data. This, however, comes with extra overhead. When granularity is tuned fine enough to the scale of single example, cross-validation would degenerate itself into a brute force method, which may exhibit intractability for applications sensitive to speed. Other ways of countering concept drift include the sliding window method [17], which maintains a sliding window with either fixed or adaptively adjustable size to determine the timeframe of the knowledge that should be reserved, and the fading factor method [18], which assigns a time-decaying factor (usually in the form of inverse exponential) to each hypothesis built over time. In this manner, old knowledge would gradually be obsoleted and could be removed when the corresponding factor downgrades to below the threshold.

Despite the popularity of the data stream study, learning from nonstationary data streams with imbalanced class distribution is a relatively uncharted area in which the difficulty lies in the context. In static context, the counterpart of this problem is recognized as "imbalanced learning," which corresponds to domains where certain types of data distribution over-dominate the instance space compared to other data distributions [19]. It is an evolving area and has attracted significant attention in the community [20–24]. However, the solutions become rather limited when imbalanced learning is set in the context of data streams. A rather straightforward way is to apply off-the-shelve imbalanced learning methods to over-sample the minority class examples in each data chunk. Following this idea, Ditzler and Chawla [25] used the synthetic minority over-sampling technique (SMOTE) [21] to create synthetic minority class instances in each data chunk arriving over time, and then applied the typical Learn^{++} framework [8] to learn from the balanced data chunks. A different way to compensate the imbalanced class ratio within each data chunk is to directly introduce the previous minority class examples into the current training data chunk. In Literature [26], *all* previous minority class examples are accommodated into the current training data chunk, upon which an ensemble of hypotheses is then built to make predictions on the datasets under evaluation. Considering the evolution of class concepts over time, an obvious heuristic to improve this idea is to only accommodate previous minority class examples that are *most similar* to the minority class set in the current training data chunk. Selectively recursive approach (SERA) [27], multiple SERA (MuSeRA) [28], and reconstruct–evolve–average (REA) [29] all stem from this idea; however, they differ in their ways of creating single or ensemble hypothesis as well as how to measure the similarity between previous and current minority class examples.

This chapter focuses mainly on introducing these three types of algorithms for learning from nonstationary data streams with imbalanced class distribution. Section 7.2 gives the preliminaries concerning different algorithms and compares different strategies of augmenting the minority class examples in training data chunks. Section 7.3 presents the algorithmic procedures of these algorithms and elaborates their theoretical foundation. Section 7.4 evaluates the efficacy of these algorithms against both real-world and synthetic benchmarks, where the type of concept drifts, the severity of imbalanced ratio, and the level of noise are all customizable to facilitate a comprehensive comparison. Section 7.5 concludes the chapter and lists several potential directions that can be pursued in the future.

7.2 PRELIMINARIES

The problem of learning from nonstationary data streams with imbalanced class distribution manifests largely in two subproblems: (i) How can imbalanced class distributions be managed? (ii) How can concept drifts be managed? An algorithm should thus be carefully designed in a way that the two subproblems could be effectively solved simultaneously. This section introduces preliminary knowledge that existing methods use to deal with these problems for a better understanding of algorithm parts described in the next section.

7.2.1 How to Manage Imbalanced Class Distribution

Increasing the number of minority class instances within the training data chunk to compensate the imbalanced class ratio is a natural way to improve prediction accuracy on minority classes. In the context of learning from a static dataset, it is recognized as an "over-sampling" method, which typically creates a set of synthetic minority class instances based on the existing ones. SMOTE [21] is the most well known in this category. Using SMOTE, a synthetic instance x is created using a random segment of the line linking a minority class example x_i and one of its k-nearest minority class neighbors, $\hat{x}_i$, that is,

$$x = x_i + \sigma \times (\hat{x}_i - x_i) \tag{7.1}$$

where $\sigma \in (0, 1)$.

Treating the arrived data chunk as a static dataset, over-sampling method can be used to augment the minority class examples therein. There is a potential flaw though that information on the minority class carried by previous data chunks cannot benefit the learning process on the current data chunk, which could result in "catastrophic forgetting" [14] for learning minority class concepts. One way to work around this is to buffer minority class examples over time and put *all* of them in the current training data chunk to compensate the imbalanced class ratio [26]. Nonetheless, in light of the concept drifts, accommodating previous minority class examples with severely deviated target concept could undermine the learning performance on minority class examples in the current training data chunk. This gives rise to the method of *selectively* accommodating previous minority class examples with the most similar target concept as current minority class into the current training data chunk.

A direct way to compare the similarity between a previous minority class example and the current minority class is to calculate the Mahalanobis distance between them [27, 28]. It differs from Euclidean distance in that it takes into account the correlations of the dataset and is scale-invariant. The Mahalanobis distance Ω from a set of n-variate instances with a mean value $\mu = [\mu_1, \ldots, \mu_n]^T$ and covariance matrix Σ to an arbitrary instance $x = [x_1, \ldots, x_n]^T$ is defined as [30]:

$$\Omega = \sqrt{(x - \mu)^T \Sigma^{-1} (x - \mu)} \tag{7.2}$$

This, however, may exhibit a potential flaw: it assumes that there are no disjoint subconcepts within the minority class concept. Otherwise, there may exist several subconcepts for the minority class, that is, Δ_1 and Δ_2 in Figure 7.1b instead of Δ in Figure 7.1a. This could be potentially improved by adopting the *k-nearest neighbors* paradigm to estimate the degree of similarity [29]. Specifically, each previous minority class example determines the number of minority examples that are within its k-nearest neighbors in the current training data chunk as its degree of similarity to the current minority class set. It can be illustrated from Figure 7.1c. Here, highlighted areas surrounded by dashed circles represent the k-nearest neighbor search area for each previous minority class example: S_1, S_2, S_3, S_4, and S_5. Search area of S_i represents the region where the k-nearest neighbors of S_i in the *current training data chunk* fall, which consists of both the majority class examples and the minority class examples. Since the majority class examples do not affect the similarity estimation, they are not shown in Figure 7.1. Current minority class examples are represented by bold circles, and the numbers of these falling in each of the "search areas" are 3, 1, 2, 1, and 0, respectively. Therefore, the similarity of S_1, S_2, S_3, S_4, and S_5 to the current minority example set is sorted as $S_1 > S_3 > S_2 = S_4 > S5$.

Using previous minority class examples to compensate the imbalanced class ratio could potentially violate the one-pass constraint [3], which mandates that previous data can never be accessed by the learning process on current training data chunk. The reason for imposing one-pass constraint for incremental learning is to avoid overflow of the limited memory due to the retention of vast amount of streaming data therein. However, given the unique nature of imbalanced learning that minority class examples are quite scarce within each training data chunk, the memory for keeping them around would be affordable.

7.2.2 How to Manage Concept Drifts

Concept drifts could be handled by solely relying on the current training data chunk. It makes sense as the current training data chunk stands for accurate

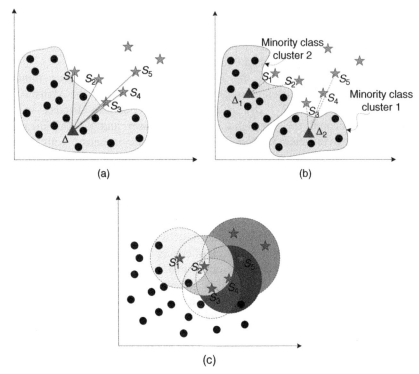

Figure 7.1 The selective accommodation mechanism: circles denote current minority class examples and stars represent previous minority class examples. (a) Estimating similarities based on distance calculation, (b) potential dilemma by estimating distance, and (c) using the number of current minority class cases within k-nearest neighbors of each previous minority example to estimate similarities.

knowledge of the class concept. However, sole maintenance of a hypothesis or hypotheses on the current training data chunk is more or less equal to discarding a significant part of previous knowledge, as knowledge of previous data chunks can never be accessed again either explicitly or implicitly once they have been processed.

One solution for addressing this issue is to maintain all hypotheses built on training data chunks over time and apply all of them to make predictions on datasets under evaluation. Concerning the way in which hypotheses combine, Gao et al. [31] employed a uniform voting mechanism to combine the hypotheses maintained this way as it is claimed that in practice the class concepts of datasets under evaluation may not necessarily evolve consistently with the streaming training data chunks. Putting aside this debatable subject, most work still assumes that the class distribution of the datasets under evaluation remains tuned to the evolution of the training data chunks. Wang et al. [32] weighed hypotheses according to their classification accuracy on *current training data chunk*. The weighted combination of all hypotheses then makes the prediction on the current dataset under evaluation. Polikar et al. [8] adopted a similar strategy to combine hypotheses in a weighted manner. The difference is the so-called "ensemble-of-ensemble" paradigm applied to create *multiple* hypotheses over each training data chunk.

7.3 ALGORITHMS

On the basis of how different algorithms compensate imbalanced class ratio of the training data chunk under consideration, this section introduces three types of algorithms, that is, "over-sampling" algorithm, "take-in-all accommodation" algorithm, and "selective accommodation" algorithm. The general incremental learning scenario is that a training data chunk $\mathcal{S}^{(t)}$ with labeled examples and a testing dataset $\mathcal{T}^{(t)}$ with unlabeled instances always come in a pair-wise manner at any timestamp t. The task of algorithms at any timestamp t is to make predictions on $\mathcal{T}^{(t)}$ as accurately as possible, based on the knowledge they have on $\mathcal{S}^t$ or the whole data stream $\{\mathcal{S}^{(1)}, \mathcal{S}^{(2)}, \ldots, \mathcal{S}^{(t)}\}$. Without loss of generality, it is assumed that the imbalanced class ratio of all training data chunks is the same. It would be easily generalized to the case when the training data chunks indeed have different imbalanced class ratios.

7.3.1 Over-Sampling Method

The most naive implementation of this method is to apply over-sampling method to augment the minority class examples within the arrived data chunk. After that, a standard classification algorithm is used to learn from the augmented training data chunk. Using SMOTE as the over-sampling technique, the implementation is shown in Algorithm 7.1.[1]

As in Algorithm 7.1, SMOTE is applied to create a synthetic minority class instance set $\mathcal{M}^{(t)}$ on top of the minority class example set $\mathcal{P}^{(t)}$ of each training data chunk $\mathcal{S}^{(t)}$. $\mathcal{M}^t$ is then appended to $\mathcal{S}^{(t)}$ to increase the class ratio of the minority class data therein, which is then used to create the final hypothesis $h_{final}^{(t)}$ to make predictions on the current testing dataset $\mathcal{T}^{(t)}$.

A more complex way to implement this idea is to create multiple hypotheses on over-sampled training data chunks and use the weighted combination of these hypotheses to make predictions on the dataset under evaluation. Ditzler and Chawla [25] followed this idea by applying the Learn^{++} paradigm, which is shown in Algorithm 7.2.

The hypotheses created on over-sampled data chunks are kept in memory over time. Whenever there is a new chunk of training data $\mathcal{S}^{(t)}$ that arrives, the algorithm first applies the SMOTE method to augment $\mathcal{S}^{(t)}$ with the synthetic minority class instances in $\mathcal{M}^{(t)}$. Then, a base hypothesis h_t is created on the

---

[1]/ $\star \ldots \star$ / represents the inline comments; the same comment will occur only once for all algorithms listed hereafter.

Algorithm 7.1   Naive implementation of the over-sampling method

---

Inputs:

 1: timestamp: t

 2: current training data chunk: $\mathcal{S}^{(t)} = \{(x_1, y_1), \ldots, (x_m, y_m)\}$

    /* $x_i \in \mathcal{X}$, $y_i \in Y$ */

 3: current data set under evaluation: $\mathcal{T}^{(t)} = \{x'_1, \ldots, x'_n\}$

    /* x'_j is the j-th instance. Class label of instances in $\mathcal{T}^{(t)}$ is unknown. */

 4: base classifier: L

    /* e.g., CART, MLP, etc., */

Procedure:

 5: **for** $t : 1 \rightarrow \ldots$ **do**

 6:    $\mathcal{S}^{(t)} \rightarrow \{\mathcal{P}^{(t)}, \mathcal{N}^{(t)}\}$

    /* $\mathcal{P}^{(t)}, \mathcal{N}^{(t)}$ are the minority and majority class sets for $\mathcal{S}^{(t)}$, respectively. */

 7:    $\mathcal{M}^{(t)} \leftarrow \text{SMOTE}(\mathcal{P}^{(t)})$

    /* class labels of instances within $\mathcal{M}$ are all minority class label. */

 8:    $h^{(t)}_{final} \leftarrow L(\{\mathcal{S}^{(t)}, \mathcal{M}^{(t)}\})$

    /* $h^{(t)}_{final} : Y \leftarrow \mathcal{X}$ */

 9:    **return** hypothesis $h^{(t)}_{final}$ for predicting the class label of any instance x' in $\mathcal{T}^{(t)}$.

---

augmented data chunk $\{\mathcal{S}^{(t)}, \mathcal{M}^{(t)}\}$. The learning performance of h_t is examined by evaluating its prediction error rate ε on the *original* $\mathcal{S}^{(t)}$. If $\varepsilon > 0.5$, that is, worse than the random guess, h_t is abandoned and the same procedure is applied again until a qualified hypothesis could be obtained. This, however, exposes a potential problem of the algorithm that steps 9–12 might be repeated many times before it can work out a hypothesis with acceptable performance, which prohibits itself from being applied to high-speed data streams. A similar procedure also applies to all base hypotheses created on previous data chunks, that is, $\{h_1, h_2, \ldots, h_{t-1}\}$. Those that fail the test, that is, $\varepsilon > 0.5$ on the current data chunk, will have their weights set to be 0. In this way, it is guaranteed that only base hypotheses that achieve satisfying performance on current training data chunks are used to constitute the ensemble classifier $h^{(t)}_{\text{final}}$ for prediction of unlabeled instances in the testing dataset $\mathcal{T}^{(t)}$.

7.3.2   Take-In-All Accommodation Algorithm

Instead of creating synthetic minority class instances to balance the training data chunk, this kind of algorithm keeps around all previous minority class examples over time and pushes them into the current training data chunk to compensate the imbalanced class ratio. An implementation of this algorithm [26] is shown in Algorithm 7.3.

   All minority class examples are kept inside the data queue $\mathcal{Q}$. Upon arrival of a new data chunk, all previous minority class examples are pushed into it to compensate its imbalanced class distribution. Then an ensemble of classifiers is

---

Algorithm 7.2 Learn^{++} for nonstationary data stream with imbalanced class distribution

---

Inputs:
1: timestamp: t
2: current training data chunk: $S^{(t)} = \{(x_1, y_1), \ldots, (x_m, y_m)\}$
3: current data set under evaluation: $T^{(t)} = \{x'_1, \ldots, x'_n\}$
4: soft-typed base classifier: L
   /* a soft-typed base classifier outputs the likelihood f that an instance x belongs to each class instead of a hard class label. */
5: hypotheses set H: $\{h_1, \ldots, h_{t-1}\}$
   /* h_i is the base hypothesis created at the data chunk with timestamp i */

Procedure:
6: **for** $t : 1 \to \ldots$ **do**
7:    $S^{(t)} \to \{\mathcal{P}^{(t)}, \mathcal{N}^{(t)}\}$
8:    **repeat**
9:       $\mathcal{M}^{(t)} \leftarrow \text{SMOTE}(\mathcal{P}^{(t)})$
10:      $h_t \leftarrow L(\{S^{(t)}, \mathcal{M}^{(t)}\})$
11:      $\varepsilon = \frac{1}{m} \times \sum_{i=1}^{m} [\![h_t(x_i) \neq y_i]\!]$
         /* $[\![c]\!] = 1$ *if and only if* c is *true* and 0 otherwise */
12:    **until** $\varepsilon < 0.5$
13:    $H \leftarrow \{H, h_t\}$
14:    $W \leftarrow \{\}$
15:    **for** $i : 1 \to t$ **do**
16:      $\varepsilon_i \leftarrow \sum_{j=1}^{m} [\![h_i(x_j) \neq y_j]\!]/m$
17:      $w_i \leftarrow \log((1 - \varepsilon_i)/\varepsilon_i)$
18:      **if** $w_i > 0$ **then**
19:         $W \leftarrow \{W, w_i\}$
20:      **else**
21:         $W \leftarrow \{W, 0\}$
22:    **return** Composite hypothesis $h_{final}^{(t)}$ for predicting the class label of $T^{(t)}$ is $H \times W$, i.e.,

$$h_{final}^{(t)}(x'_j) = \arg\max_{c \in Y} \sum_{i=1}^{t} w_i \times f_i^c(x'_j) \tag{7.3}$$

/* $f_i^c(x_j)$ is the *a posteriori* probability for x'_j belonging to class c output by h_i. */

---

created on the augmented data chunk to make predictions on the testing dataset. The method of creating the ensemble classifiers is similar to the classic bagging method [33] in that multiple hypotheses created on randomly permutated sets are aggregated with uniform weights into a composite classifier. However, considering the limited number of minority class examples available in training data chunk, only the majority class set is randomly sampled without replacement. This differs from classic bagging in that the sample sets have no overlaps at all. Therefore, it is referred as *uncorrelated bagging* (UB) in the rest of this chapter. The dataset that is used to create base hypothesis h contains exactly the same minority examples and different majority class examples. In this way,

---

Algorithm 7.3 Uncorrelated bagging with take-in-all accommodation of previous minority class examples

---

Inputs:
1: timestamp: t
2: current training data chunk: $\mathcal{S}^{(t)} = \{(x_1, y_1), \ldots, (x_m, y_m)\}$
3: current data set under evaluation: $\mathcal{T}^{(t)} = \{x'_1, \ldots, x'_n\}$
4: minority class data queue: $\mathcal{Q}$
    /* $\mathcal{Q}$ stores all previous minority class examples before the current time t. */
5: soft-typed base classifier: L

Procedure:
6: **for** $t : 1 \to \ldots$ **do**
7:    $\mathcal{S}^{(t)} \leftarrow \{\mathcal{P}^{(t)}, \mathcal{N}^{(t)}\}$
       /* Assume $||\mathcal{P}^{(t)}|| = p$ and $||\mathcal{N}^{(t)}|| = q$ */
8:    $\mathcal{P}'^{(t)} \leftarrow \{\mathcal{P}^{(t)}, \mathcal{Q}^{(t)}\}$
       /* Assume $||\mathcal{P}'^{(t)}|| = p'$, and $\frac{q}{p'} = K$ */
9:    **for** $k \leftarrow 1 \ldots K$ **do**
10:       $\mathcal{N}'^{(k)} \leftarrow$ *sample_without_replacement*$(N^{(t)})$
          /* $||\mathcal{Q}'^{(k)}|| = p'$ */
11:       $h_k^{(t)} \leftarrow L(\{\mathcal{Q}'^{(k)}, \mathcal{P}'^{(k)}\})$
12:    $\mathcal{Q} \leftarrow \{\mathcal{Q}, \mathcal{P}\}$
13:    **return** Averaged composite hypothesis $h_{final}^{(t)}$ for predicting class label of any instance x'_j within $\mathcal{T}^{(t)}$:

$$h_{final}^{(t)}(x'_j) = \arg\max_{c \in Y} \frac{1}{k} \times \sum_{i=1}^{K} f_i^c(x'_j) \tag{7.4}$$

---

the best use could be made of the minority class examples, while minimizing the correlation of different base hypotheses. This is particularly important for designing a decent composite classifier as *diversity* across base hypotheses plays a crucial role in lifting the performance of aggregating ensemble of hypotheses as compared to that of a single one. Finally, the minority class examples within the current data chunk are pushed back to $\mathcal{Q}$ to facilitate imbalanced learning on future data chunks.

The independence across base hypotheses can help reduce the overall error rate of prediction. Assuming that the estimated *a posteriori* probability of base hypothesis h_k for an instance x is $f^k(x)$, the output of the composite classifier h_{final} for x is,

$$f^E(x) = \frac{1}{K} \sum_{k=1}^{K} f^k(x) \tag{7.5}$$

Figure 7.2 shows the error regions introduced by estimating the *true* Bayes model, in which $\mathcal{P}_{i/j}$ and $f_{i/j}$ represent the *a posteriori* probability of the outputs of classes i and j by true Bayes model and estimated Bayes model, respectively. x^* and x^b are where the true and estimated Bayes models output exactly the

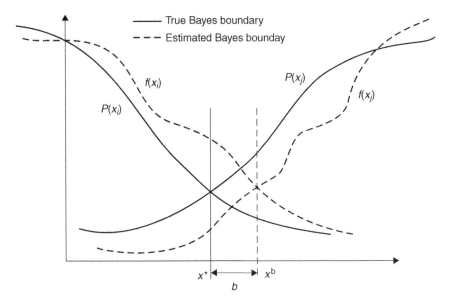

Figure 7.2 Error regions associated with approximating the *a posteriori* probability.

same *a posteriori* probability for classes i and j, that is,

$$P(x^*|i) = P(x^*|j) \tag{7.6}$$

$$f_i(x^b) = f_j(x^b) \tag{7.7}$$

Tumer and Ghosh [34] proved that classification error is proportional to the boundary error $b = x^b - x^*$. According to Equation 7.6,

$$f_i(x^* + b) = f_j(x^* + b) \tag{7.8}$$

Tumer and Ghosh [34, 35] showed that the output of any Bayes classifier can be decomposed into the true Bayes output plus error, that is,

$$f_c(x) = P(x|c) + \beta_c + \eta_c(x) \tag{7.9}$$

where β_c and η_c are the *bias* and *variance* of the hypothesis.
According to Equation 7.9, Equation 7.8 can be rewritten as

$$P(x^* + b|i) + \beta_i + \eta_i(x^b) = P(x^* + b|j) + \beta_j + \eta_j(x^b) \tag{7.10}$$

Equation 7.10 can be further approximated by Taylor theorem, that is,

$$P(x^*|i) + b \times P'(x^*|i) + \beta_i + \eta_i(x^b) \simeq P(x^*|j) + b \times P'(x^*|j) + \beta_j + \eta_j(x^b) \tag{7.11}$$

According to Equation 7.6, $\mathcal{P}(\boldsymbol{x}^*|i)$ and $\mathcal{P}(\boldsymbol{x}^*|j)$ can be canceled from the equation; thus,

$$b \times \mathcal{P}'(\boldsymbol{x}^*|i) + \beta_i + \eta_i(\boldsymbol{x}^b) = b \times \mathcal{P}'(\boldsymbol{x}^*|j) + \beta_j + \eta_j(\boldsymbol{x}^b) \qquad (7.12)$$

which gives the equation for calculating boundary error as

$$b = \frac{1}{\mathcal{P}'(\boldsymbol{x}^*|i) - \mathcal{P}'(\boldsymbol{x}^*|j)} \times [(\beta_j - \beta_i) + (\eta_j(\boldsymbol{x}^b) - \eta_i(\boldsymbol{x}^b))] \qquad (7.13)$$

where $\mathcal{P}'(\boldsymbol{x}^*|i) - \mathcal{P}'(\boldsymbol{x}^*|j)$ only concerns the true Bayes model, and is thus independent of any classifier. This means,

$$b \propto (\beta_j - \beta_i) + (\eta_j(\boldsymbol{x}^b) - \eta_i(\boldsymbol{x}^b)) \qquad (7.14)$$

if $\eta_c(\boldsymbol{x})$ is identically distributed and has a Gaussian distribution with mean 0 and variance $\sigma_{\eta_c}^2$; b thus has a Gaussian distribution with mean $\dfrac{\beta_j - \beta_i}{\mathcal{P}'(\boldsymbol{x}^*|i) - \mathcal{P}'(\boldsymbol{x}^*|j)}$ and variance as

$$\sigma_b^2 = \frac{\sigma_{\eta_i}^2 + \sigma_{\eta-j}^2}{(\mathcal{P}'(\boldsymbol{x}^*|i) - \mathcal{P}'(\boldsymbol{x}^*|j))^2} \qquad (7.15)$$

According to Equation 7.9, the output of the composite classifier can be decomposed into

$$f_c^E \boldsymbol{x} = \mathcal{P}(\boldsymbol{x}|c) + \overline{\beta}_c + \overline{\eta}_c(\boldsymbol{x}) \qquad (7.16)$$

Assuming that the errors of individual base classifiers are independent of each other, the variance of $\overline{\eta}_c(\boldsymbol{x})$ is as follows [26]:

$$\sigma_{\overline{\eta}_c}^2 = \frac{1}{k^2}\sigma_{\eta_c^i}^2 \qquad (7.17)$$

Therefore, based on Equation 7.15, the variance of the boundary error of the composite hypothesis can be significantly reduced by a factor of $1/k^2$ as compared to the base classifier, that is,

$$\sigma_{bE}^2 = \frac{1}{k^2}\sigma_{b^i}^2 \qquad (7.18)$$

7.3.3 Selective Accommodation Algorithm

In lieu of pushing all minority class examples into the data chunk under consideration, this type of method evaluates the similarity between each minority class example and the current minority class set first. Only parts of the previous minority class examples that are most similar to the target concept of the current minority class are chosen to compensate the imbalanced class ratio of the data chunk under consideration. As discussed in Section 7.2.1, the similarity can be calculated by measuring the Mahalanobis distance between the previous minority class example and the current minority class set or counting the number of minority class cases within the k-nearest neighbors of each previous minority class example inside the data chunk under consideration. The Mahalanobis distance approach is adopted by SERA and MuSeRA, while the k-nearest neighbor approach is applied in REA. The pseudo-codes of these three algorithms are described in Algorithm 7.4.

The system-level framework of SERA, MuSeRA, and REA is illustrated in Figure 7.3. All previous minority class examples are stored in $\mathcal{Q}$. At time $t = n$, a certain number of minority class examples in $\mathcal{Q}$ are chosen on the basis of alternative similarity measurement, that is, the Mahalanobis distance or the k-nearest neighbors. These examples are then appended to $S^{(n)}$ such that the ratio of minority class examples in the "post-balanced" training data chunk $\{S^{(n)}, \mathcal{Q}(I)\}$ is equal to f. Base hypothesis h_n is created on $\{S^{(n)}, \mathcal{Q}(I)\}$. SERA stops at this point to return h_n as the final hypothesis $h_{\text{final}}^{(n)}$. MuSeRA and REA continue by inserting h_n into the hypothesis queue H, each base hypothesis h_i within which

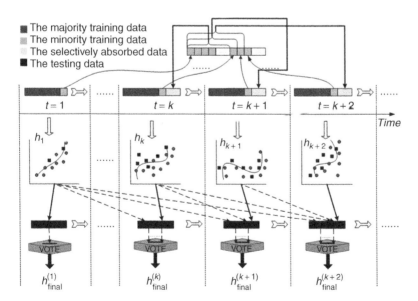

Figure 7.3 The system-level framework of SERA, MuSeRA, and REA.

Algorithm 7.4   Selective accommodation of previous minority class examples

---

Inputs:
  1: current timestamp t.
  2: current training data chunk: $\mathcal{S}^{(t)} = \{(x_1, y_1), \ldots, (x_m, y_m)\}$
  3: current data set under evaluation: $\mathcal{T}^{(t)} = \{x'_1, \ldots, x'_n\}$
  4: minority class data queue: $\mathcal{Q}$
  5: soft-typed base classifier: L
  6: selective accommodation method: d
     /* d could be *SERA*, *MuSeRA*, or *REA* */
  7: post-balanced ratio f
     /* desired minority class to majority class ratio. */
  8: hypotheses set $H = \{h_1, h_2, \ldots, h_{t-1}\}$

Procedure:
  9: **for** $t : 1 \to \ldots$ **do**
10:     $\mathcal{S}^{(t)} \leftarrow \{\mathcal{P}^{(t)}, \mathcal{N}^{(t)}\}$
      /* Assume $||\mathcal{P}^{(t)}|| = p$ and $||\mathcal{N}^{(t)}|| = q$ */
11:     **if** $(||\mathcal{P}^{(t)}|| + ||\mathcal{Q}||)/||\mathcal{S}^{(t)}|| <= f$ **then**
12:        $h_t \leftarrow L(\{\mathcal{S}^{(t)}, \mathcal{Q}\})$
13:     **else**
14:        $n \leftarrow f \times ||\mathcal{S}^t|| - ||\mathcal{P}^t||$
15:        $m \leftarrow \{\}$
16:        **for** $x_j \in \mathcal{Q}$ **do**
17:           **if** $d = REA$ **then**
18:              $\mathcal{K} \leftarrow$ *k-nearest-neighbor*$(x_j, \mathcal{S}^{(t)})$
              /* calculates the k-nearest neighbors of x_j within $\mathcal{S}^{(t)}$ */
19:              $m \leftarrow \{m, \sum_{k \in \mathcal{K}} [\![k \in \mathcal{P}^{(t)}]\!]\}$
              /* number of minority cases within the k-nearest neighbor of x_j */
20:           **else**
21:              $\omega = \sqrt{(x_j - \mu)^T \Sigma^{-1} (x_j - \mu)}$
              /* μ and Σ are the mean and the covariance matrix of $\mathcal{P}^t$ */
22:              $m \leftarrow \{m, 1/\omega\}$
23:        $(m', I) \leftarrow$ *reverse-sort*(m)
          /* sort m in descending order, and put the corresponding indices in I */
24:        $I \leftarrow I(1 : n)$
          /* Use the most similar previous minority class examples to augment $\mathcal{S}^{(t)}$ to achieve desired class ratio f. */
25:        $h_t \leftarrow L(\mathcal{S}^t + \mathcal{Q}^t(I))$
26:     $\mathcal{Q} \leftarrow \{\mathcal{Q}, \mathcal{P}^t\}$
27:     **if** $d = SERA$ **then**
28:        **return** $h^t_{final} = h_t$ for predicting class label of instance x' within $\mathcal{T}^{(t)}$
29:     **else**
30:        $H \leftarrow \{H, h_t\}$
31:        $W \leftarrow \{\}$
32:        **for** $i : 1 \to t$ **do**
33:           $e_i = \frac{1}{|\mathcal{S}_t|} \sum_{(x_j, y_j) \in \mathcal{S}_t} (1 - f_i^{y_j}(x_j))^2$
34:           $W \leftarrow \{W, \log 1/e_i\}$
35:        **return** Composite hypothesis $h^{(t)}_{final}$ for predicting class label of any instance x'_j in testing data set $\mathcal{T}_t$ is $H \times W$, i.e.,

$$h^{(t)}_{final}(x'_j) = \arg \max_{c \in Y} \sum_{i=1}^{t} w_i \times f_i^c(x'_j) \qquad (7.19)$$

is applied onto $S^{(n)}$ to calculate the error rate e_i. The reverse logarithm of e_i is then used as the weight w_i for each base hypothesis h_i. Since w_i is inversely proportional to the error rate e_i of h_i, a strong hypothesis automatically gets higher votes in constituting the final hypothesis h_{final}^n. One should be cautious in dealing with the situation when w_i is either too small or too large, which would mean either an obsoleted or an over-fitting h_i.

In theory, MuSERA/REA can obtain a smaller error bound than that of "UB" introduced in Section 7.3.2.

By making the same assumption that the Bayes error rate comes from the *variance* η_c^i of the base hypothesis h_i as for the analysis of UB, as weights $\{w_i\}$ are inversely proportional to the error rates $\{e_i\}$, they can be approximated by

$$w_i = \frac{C}{\sigma_{\eta_c^i}^2} \tag{7.20}$$

where $\sigma_{\eta_c^i}^2$ is the variance of η_c^i and C is a constant for all base hypotheses $\{h_i\}$.

On the basis of Equation 7.9, η_c^i is a part of the probability output; thus, the *variance* η_C^E of the final hypothesis at time stamp n is

$$\eta_C^E(x) = \frac{\sum_{i=1}^n w_i \eta_c^i(x)}{\sum_{i=1}^n w_i} \tag{7.21}$$

The variance of η_C^E is

$$\sigma_{\eta_C^E}^2 = \frac{\sum_{i=1}^n w_i^2 \sigma_{\eta_c^i}^2}{\sum_{i=1}^n w_i^2} \tag{7.22}$$

According to Equation 7.20, Equation 7.22 can be simplified into

$$\sigma_{\eta_C^E}^2 = \frac{1}{\sum_{i=1}^n 1/\sigma_{\eta_C^i}^2} \tag{7.23}$$

On the other hand, there is Inequation 7.24:

$$\sum_{i=1}^n \sigma_{\eta_C^i}^2 \times \sum_{i=1}^n \frac{1}{\sigma_{\eta_C^i}^2} \geq n^2 \tag{7.24}$$

The proof is straightforward, assuming $\sigma_{\eta_C^i}^2 = a_i$:

$$\sum_{i=1}^n a_i \times \sum_{j=1}^n \frac{1}{a_j} = \sum_{\substack{i=1,j=1 \\ i=j}}^n a_i \times \frac{1}{a_j} + \sum_{\substack{i=1,j=1 \\ i\neq j}}^n a_i \times \frac{1}{a_j} \tag{7.25}$$

$$= n + \frac{1}{2} \sum_{\substack{i=1, j=1 \\ i \neq j}}^{n} \left(\frac{a_i^2 + a_j^2}{a_i a_j} \right)$$

$$\geq n + \binom{n}{2} \times 2 = n + n^2 - n = n^2$$

Therefore, according to Equation 7.23 and Inequation 7.24,

$$\sigma_{\eta_C^E}^2 \leq \frac{1}{n^2} \sum_{i=1}^{n} \sigma_{\eta_C^i}^2 \tag{7.26}$$

The single classifier developed by MuSeRA/REA is based on the big picture of the training data chunk augmented by the previous minority class examples, while the single classifier developed by UB is based on a partial view. Thus, it is valid that the error rate of the base hypothesis h_i created by MuSeRA/REA should be equal to or less than that created by UB, that is,

$$\sigma_{\eta_C^i}^2 \leq \sigma_{b^i}^2 \tag{7.27}$$

which naturally leads to the same result regarding the comparison between their averaged error rates, that is,

$$\frac{1}{n} \sum_{i=1}^{n} \sigma_{\eta_C^i}^2 \leq \frac{1}{k} \sum_{j=1}^{k} \sigma_{b^i}^2 \tag{7.28}$$

Given the arrival of sufficient training data chunks, it is inevitable that $n > k$; thus,

$$\frac{1}{n^2} \sum_{i=1}^{n} \sigma_{\eta_C^i}^2 \leq \frac{1}{k^2} \sum_{j=1}^{k^2} \sigma_{b^i}^2 \tag{7.29}$$

On the basis of Equations 7.18, 7.26, and 7.29, there is

$$\sigma_{\eta_C^E}^2 \leq \sigma_{b^E}^2 \tag{7.30}$$

which proves that MuSeRA/REA can provide less erroneous prediction results than UB.

7.4 SIMULATION

To provide a more comprehensive insight into the algorithms introduced in Section 7.2, simulations are conducted to compare their performance against both synthetic and real-world benchmark datasets, and are listed and configured as follows.

- The *REA* algorithm uses k-nearest neighbors to determine the qualification of previous minority class examples for making up the minority class ratio f in post-balanced training data chunks. k is set to be 10 and f is set to be 0.5.
- The *SERA* algorithm uses Mahalanobis distance to determine the qualification of previous minority class examples for making up the minority class ratio f in post-balanced training data chunks. f is set to be 0.5.
- The *UB* algorithm uses all previous minority class examples to balance the training data chunk.
- The *SMOTE* algorithm employs the synthetic minority over-sampling technique to create a number of synthetic minority class instances for making up the minority class ratio f in post-balanced training data chunk. f is set to be 0.5.
- The *Normal* approach directly learns on the training data chunk; it is the baseline of this simulation.

7.4.1 Metrics

Following the routine of imbalanced learning study, the minority class data and the majority class data belong to positive and negative classes, respectively. Let $\{p, n\}$ denote the positive and negative true class labels and $\{Y, N\}$ denote the predicted positive and negative class labels; the confusion matrix for the binary classification problem can be defined as in Figure 7.4.

By manipulating the confusion matrix, the overall prediction accuracy (OA) can be defined as

$$OA = \frac{TP + TN}{TP + TN + FP + FN} \tag{7.31}$$

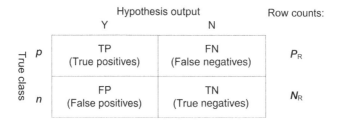

Figure 7.4 Confusion matrix for binary classification.

OA is usually adopted in the traditional learning scenario, that is, static datasets with balanced class distribution, to evaluate the performance of algorithms. However, when the context changes to imbalanced learning, it is wise to apply other metrics for such evaluation [19], among which receiver operation characteristics (ROC) curve and area under ROC curve (AUROC) are the most strongly recommended [36].

On the basis of the confusion matrix as defined in Figure 7.4, one can calculate the TP_rate and FP_rate as follows:

$$\text{TP\_rate} = \frac{\text{TP}}{P_R} = \frac{\text{TP}}{\text{TP} + \text{FN}} \tag{7.32}$$

$$\text{FP\_rate} = \frac{\text{FP}}{N_R} = \frac{\text{FP}}{\text{FP} + \text{TN}} \tag{7.33}$$

ROC space is established by plotting TP_rate over FP_rate. Generally speaking, hard-type classifiers (those that output only discrete class labels) correspond to points in ROC space (FP_rate, TP_rate). On the other hand, soft-type classifiers (those that output a likelihood that an instance belongs to either class label) correspond to curves in ROC space. Such curves are formulated by adjusting the decision threshold to generate a series of points in ROC space. For example, if the likelihoods of an unlabeled instance x_k belonging to minority class and majority class are 0.3 and 0.7, respectively, natural decision threshold $d = 0.5$ would classify x_k as a majority class example as $0.3 < d$. However, d could also be set to other values, for example, $d = 0.2$. In this case, x_k would be classified as a minority class example as $0.3 > d$. By tuning d from 0 to 1 with a small step Θ, for example, $\Theta = 0.01$, a series of pair-wise points (FP_rate, TP_rate) could be created in ROC space. In order to assess the performance of different classifiers in this case, one generally uses AUROC as an evaluation criterion; it is defined as the area between the ROC curve and the horizontal axis (axis representing FP_rate).

In order to reflect the ROC curve characteristics for all random runs, the vertical averaging approach [36] is adopted to plot the averaged ROC curves. Implementation of the vertical averaging method is illustrated in Figure 7.5. Assume one would like to average two ROC curves, l_1 and l_2; both are formed by a series of points in the ROC space. The first step is to evenly divide the range of FP_rate into a set of intervals. Then at each interval, find the corresponding TP_rate values of each ROC curve and average them. In Figure 7.5, X_1 and Y_1 are the points from l_1 and l_2 corresponding to the interval FP_rate 1. By averaging their TP_rate values, the corresponding ROC point Z_1 on the averaged ROC curve is obtained. However, there exist some ROC curves that do not have corresponding points on certain intervals. In this case, one can use the linear interpolation method to obtain the averaged ROC points. For instance, in Figure 7.5, the point $\overline{X}$ (corresponding to FP_rate 2) is calculated on the basis of the linear interpolation of the two neighboring points X_2 and X_3. Once $\overline{X}$ is obtained,

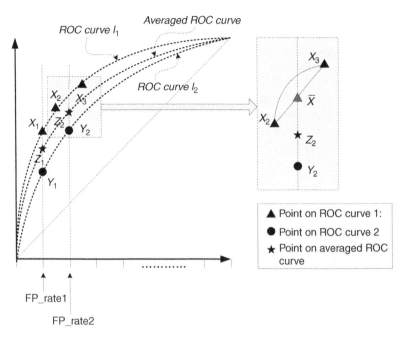

Figure 7.5 Vertical averaging approach for multiple ROC curves.

it can be averaged with Y_2 to get the corresponding Z_2 point on the averaged ROC curve.

7.4.2 Results

7.4.2.1 SEA Dataset SEA dataset [15] is a popular artificial benchmark to assess the performance of stream data-mining algorithms. It has three features randomized in [0, 10], where the class label is determined by whether the sum of the first two features surpasses a defined threshold. The third feature is irrelevant and can be considered as noise to test the robustness of the algorithm under simulation. Concept drifts are designed to adjust the threshold periodically such that the algorithm under simulation would be confronted with an *abrupt* change in class concepts after it lives with a stable concept for several data chunks.

Following the original design of the SEA dataset, the whole data streams are sliced into four blocks. Inside each of these blocks, the threshold value is fixed, that is, the class concepts are unchanged. However, whenever it comes to the end of a block, the threshold value will be changed and retained until the end of the next block. The threshold values of the four blocks are set to be 8, 9, 7, and 9.5, respectively, which again adopt the configuration of Street and Kim [15]. Each block consists of 10 data chunks, each of which has 1000 examples as the training dataset and 200 instances as the testing dataset. Examples with the sum of the two features greater than the threshold belong to the majority class, and those otherwise reside in the minority class. The number of generated minority class data is restricted to be $1/20$ of the total number of data in the corresponding data chunk. In other words, the imbalanced ratio is set to be 0.05 in our simulation. In order to introduce some uncertainty/noise into the dataset, 1% of the examples inside each training dataset are randomly sampled to reverse their class labels. In this way, approximately $1/6$ of the minority examples are erroneously labeled, which raises a challenge on handling noise for all comparative algorithms learning from this dataset.

The simulation results for the SEA dataset are averaged over 10 random runs. During each random run, the dataset is basically generated all over again. To view the performance of the algorithms in the whole learning life, "observation points" are installed on chunks 5, 10, 15, 20, 25, 30, 35, and 40.

The tendency lines of the averaged prediction accuracy over the observation points are plotted in Figure 7.6a. One can conclude from this figure that (i) REA can provide higher prediction accuracy on testing data over time than UB, which is consistent with the theoretical conclusion made in Section 7.3.3; (ii) REA does not perform superiorly in terms of OA to other comparative algorithms over time. In fact, it is the baseline ("Normal") that provides the most competitive results in terms of the OA on testing data most of the time. However, as discussed previously, OA is not of primary importance in the imbalanced learning scenario. It is metrics such as ROC/AUROC that determine how well the algorithm performs on imbalanced datasets.

The AUROC values of the comparative algorithms on the observation points are given in Figure 7.6b, complemented by the corresponding ROC curves on data chunks 10 (Fig. 7.7a), 20 (Fig. 7.7b), 30 (Fig. 7.7c), and 40 (Fig. 7.7d), respectively, as well as the corresponding numeric AUROC values on these data chunks given in Table 7.1. In this metric, REA gives superior performance over other algorithms, and SERA can generally be better than the baseline. Besides that, it is inclusive to make judgment regarding the comparison among the rest of the algorithms as well as the baseline.

7.4.2.2 ELEC Dataset The electricity market dataset (ELEC dataset) [37] is used as a real-world dataset to validate the effectiveness of the proposed algorithm in real-world applications. The data were collected from the Australian New South Wales (NSW) Electricity Market to reflect the electricity price fluctuation (Up/Down) affected by demand and supply of the market. Since the influence of the market on electricity price evolves unpredictably in the real world, the concrete representation of concept drifts embedded inside the dataset is inaccessible, which is obviously different from synthetic datasets that set up the concept drift by hand.

The original dataset contains 4531 examples dated from May 1996 to December 1998. We only retain examples after May 11, 1997, for this simulation because

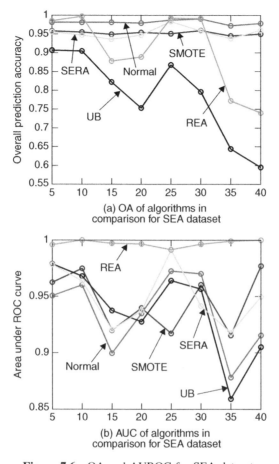

Figure 7.6   OA and AUROC for SEA dataset.

several features are missing from the examples before that date. Each example consists of eight features. Features 1–2 represent the date and the day of the week (1–7) for the collection of the example, respectively. Each example is sampled within a timeframe of 30 min, that is, a period; thus, there are altogether 48 examples collected for each day, which correspond to 48 periods a day. Feature 3 stands for the period (1–48) in which the very example was collected, and thus is a purely periodic number. Features 1–3 are excluded from the feature set as they just stand for the timestamp information of the data. According to the data sheet instruction, feature 4 should also be ignored from the learning process. Therefore, the remaining features are the NSW electricity demand, the Victoria (VIC) price, the VIC electricity demand, and the scheduled transfer between states, respectively. In summary, 27, 549 examples with the last four features are extracted from the ELEC dataset for simulation.

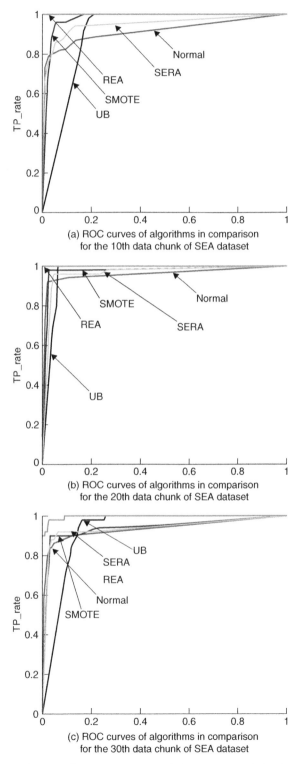

(a) ROC curves of algorithms in comparison
for the 10th data chunk of SEA dataset

(b) ROC curves of algorithms in comparison
for the 20th data chunk of SEA dataset

(c) ROC curves of algorithms in comparison
for the 30th data chunk of SEA dataset

Figure 7.7   ROC curves for selected data chunks of SEA dataset.

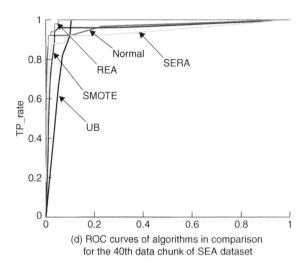

(d) ROC curves of algorithms in comparison
for the 40th data chunk of SEA dataset

Figure 7.7   (*Continued*)

Table 7.1   AUROC Values for Selected Data Chunks of SEA Dataset

Data Chunk	Normal	SMOTE	UB	SERA	REA
10	0.9600	0.9749	0.9681	0.9637	**1.0000**
20	0.9349	0.9397	0.9276	0.9373	**0.9966**
30	0.9702	0.9602	0.9565	0.9415	**0.9964**
40	0.9154	0.9770	0.9051	0.9497	**1.0000**

Bold values represents the highest AUROC values measured at running algorithms in comparison against 10th, 20th, 30th, 40th data chunk of SEA and ELEC datasets, respectively.

With the order of the examples unchanged, the extracted dataset is evenly sliced into 40 data chunks. Inside each data chunk, examples that represent electricity price going down are determined as the majority class data, while the others that represent electricity price going up are randomly under-sampled as the minority class data. The imbalanced ratio is set to be 0.05, which means only 5% of the examples inside each data chunk belong to the minority class. To summarize the preparation of this dataset, 80% of the majority class data and the minority class data inside each data chunk are randomly sampled and merged as the training data, and the remaining are used to assess the performance of the corresponding trained hypotheses.

The results of the simulation are based on 10 random runs, where the randomness comes from the random under-sampling of the minority class data. As in the procedure for SEA datasets, observation points are set up in data chunks 5, 10, 15, 20, 25, 30, 35, and 40, respectively.

Figure 7.8a plots the averaged OA of the comparative algorithms. One might be misled into believing that baseline is the best algorithm because it gives a better OA rate than the algorithms after data chunk 25. This could be easily proved wrong. Assume that a dumb method is introduced, which just classifies

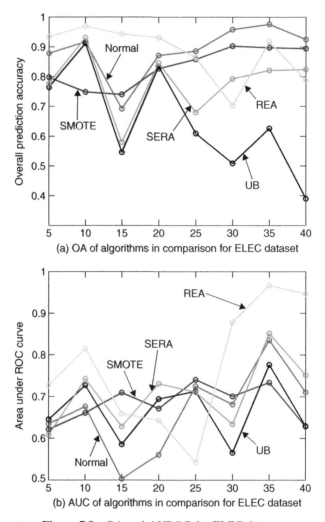

Figure 7.8   OA and AUROC for ELEC dataset.

all instances into minority class. Since the imbalanced class ratio is just 0.05, this dumb method can achieve 95% OA all the time on testing datasets, but it should never be considered in practice as it fails entirely on minority class. This fact reassures that OA is really not a good choice for validating algorithms on imbalanced datasets. The other fact that can be observed is that REA provides consistently higher OA rate than UB, which again resonates with the theoretical analysis in Section 7.3.

Figure 7.8b shows the averaged AUROC of the comparative algorithms. As complements, Figure 7.9 shows the ROC curves averaged by 10 random runs of comparative algorithms on data chunks 10 (Fig. 7.9a), 20 (Fig. 7.9b), 30

Table 7.2   AUROC Values for Selected Data Chunks of ELEC Dataset

Data Chunk	Normal	SMOTE	UB	SERA	REA
10	0.6763	0.6608	0.7273	0.7428	**0.8152**
20	0.5605	0.6715	0.6954	**0.7308**	0.6429
30	0.6814	0.7007	0.5654	0.6339	**0.8789**
40	0.7102	0.6284	0.6297	0.7516	**0.9462**

Bold values represents the highest AUROC values measured at running algorithms in comparison against 10th, 20th, 30th, 40th data chunk of SEA and ELEC datasets, respectively.

(Fig. 7.9c), and 40 (Fig. 7.9d). Table 7.2 gives the numerical values for AUROC of all comparative algorithms on selected data chunks. The data collected validate that OA is not a decisive metric for imbalanced learning, as many algorithms here can outperform the baseline. REA provides the best AUROC result over other algorithms after data chunk 25, followed by SERA.

7.4.2.3   SHP Dataset   As proposed in [32], the spinning hyperplane (SHP) dataset defines the class boundary as a hyperplane in n dimensions by coefficients $\alpha_1, \alpha_2, \ldots, \alpha_n$. An example $x = (x_1, x_2, \ldots, x_n)$ is created by randomizing each feature in the range $[0, 1]$, that is, $x_i \in [0, 1]$, $i = 1, \ldots, n$. A constant bias is defined as

$$\alpha_0 = \frac{1}{2} \sum_{i=1}^{n} \alpha_i \tag{7.34}$$

Then, the class label y of the example x is determined by

$$y = \begin{cases} 1 & \sum_{i=1}^{n} \alpha_i x_i \geq \alpha_0 \\ 0 & \sum_{i=1}^{n} \alpha_i x_i < \alpha_0 \end{cases} \tag{7.35}$$

In contrast to the *abrupt* concept drifts in SEA dataset, the SHP dataset embraces a *gradual* concept drift scheme in that the class concepts undergo a "shift" whenever a new example is created. Specifically, part of the coefficients $\alpha_1, \ldots, \alpha_n$ will be randomly sampled to have a small increment Δ added whenever a new example has been created, which is defined as

$$\Delta = s \times \frac{t}{N} \tag{7.36}$$

where t is the magnitude of change for every N example, and s alternates in $[-1, 1]$, specifying the direction of change and has a 20% chance of being reversed for every N example. α_0 is also modified thereafter using Equation 7.34. In this way, the class boundary would be similar to an SHP in the process of creating data. A dataset with gradual concept drifts requires the learning algorithm

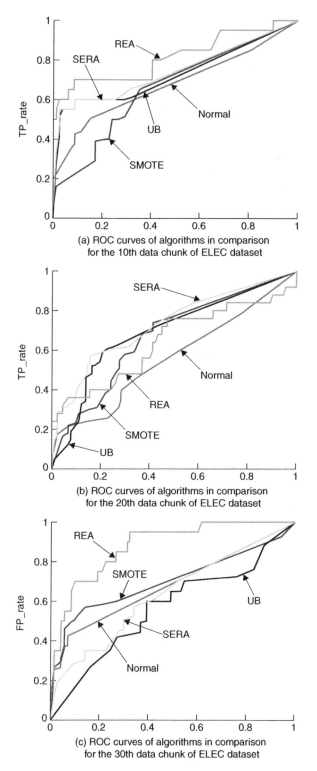

Figure 7.9 ROC curves for selected data chunks of ELEC dataset.

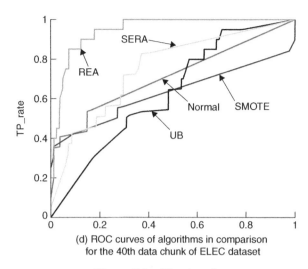

(d) ROC curves of algorithms in comparison
for the 40th data chunk of ELEC dataset

Figure 7.9   (*Continued*)

to adaptively tune its inner parameters constantly in order to catch up with the continuous updates of class concepts.

Following the procedure given in [32], the number of dimensions for the feature set of SHP dataset is deemed to be 10 and the magnitude of change t is set to be 0.1. The number of chunks is set to be 100 instead of 40 as for the previous two datasets, as we would like to investigate REA in longer series of streaming data chunks. Each data chunk has 1000 examples as the training dataset and 200 instances as the testing dataset, that is, $N = 1200$. In addition to the normal setup of the imbalanced ratio being 0.05 and noise level being 0.01, other suites of datasets are generated when the imbalanced ratio is 0.01 and noise level is 0.01, and yet another when the imbalanced ratio is 0.05 and the noise level is 0.03. In this way, the robustness of all algorithms under comparison that handle more imbalanced dataset and more noisy datasets can be evaluated. In the rest of this section, the three different setups of imbalanced ratio and noise level are referred as *setup 1* (imbalanced ratio = 0.05 and noise level = 0.01), *setup 2* (imbalanced ratio = 0.01 and noise level = 0.01), and *setup 3* (imbalanced ratio = 0.05 and noise level = 0.03), respectively.

Similar to the procedure for the previous two datasets, the results of all comparative algorithms on an SHP dataset are based on the average of 10 random runs. The observation points are placed in data chunks 10, 20, 30, 40, 50, 60, 70, 80, 90, and 100.

Figures 7.10–7.12 plot the tendency lines of OA and AUROC for comparative algorithms across observation points under *setup 1*, *setup 2*, and *setup 3*, respectively. One can see that in terms of both OA and AUROC, REA consistently outperforms all algorithms all the time. In terms of AUROC, UB and SERA compete with each other for the second place after REA in different scenarios, that is, different imbalanced ratios and different noise levels.

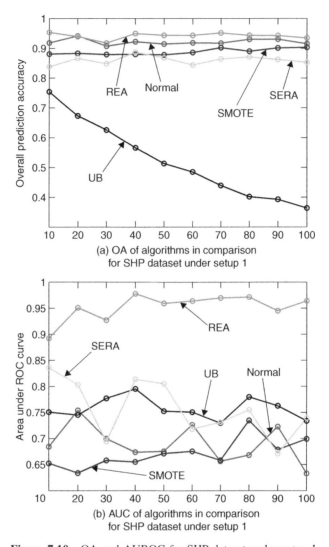

Figure 7.10 OA- and AUROC for SHP dataset under *setup 1*.

7.4.3 Hypothesis Removal

In the scenario of long-term learning from data streams, retaining all hypotheses in memory over time may not be a good option. Besides the concern for memory occupation, hypotheses built in the distant past may hinder the classification performance on current testing dataset, which should therefore somehow be pruned/removed from the hypotheses set H. This is an issue that is unique for REA, as other algorithms either just rely on the hypothesis created on the current

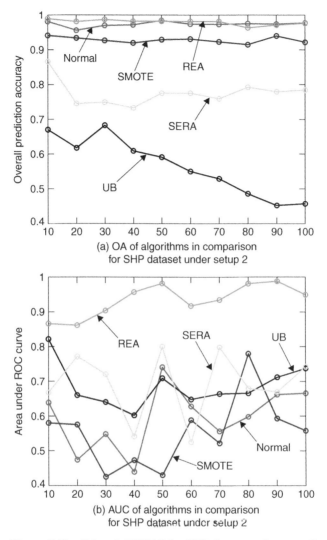

Figure 7.11 OA and AUROC for SHP dataset under *setup 2*.

training data chunk or already have a mechanism to obsolete the old hypotheses, such as Learn[++].

Imagine the hypotheses set H defined in Algorithm 7.4, which is a first-in-first-out (FIFO) queue. The original design of REA physically sets the capacity of H to be infinity, as from the time of its creation, any hypothesis will stay in memory until the end of the data stream. Now let us assume that H has a smaller capacity. Should the number of stored hypotheses exceed the capacity of H, the oldest hypothesis would automatically be removed from H. In this way, it can

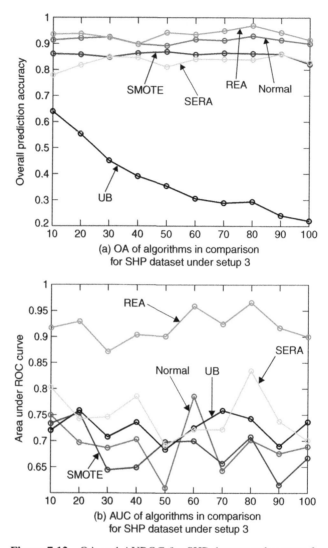

Figure 7.12   OA and AUROC for SHP dataset under *setup 3*.

be guaranteed that H always maintains the "freshest" subset of the generated hypotheses in memory.

Figure 7.13 shows the AUROC performance of REA on learning from SHP datasets with 100 data chunks when the size of H, that is, $|H|$, is equal to ∞, 60, 40, and 20, respectively. One can see that REA initially improves its performance in terms of AUROC when $|H|$ shrinks. However, when $|H| = 20$, the performance of REA deteriorates, which is worse than the case when $|H| = \infty$. On the basis of these observations, one can conclude that there exists a trade-off between $|H|$ and the REA's performance. A heuristic would be to set $|H|$

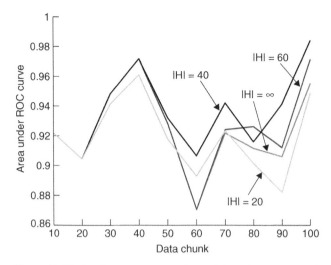

Figure 7.13 Performance comparison of hypotheses removal.

approximately half the total number of data chunks received over time, which is impractical as the number of data chunks is usually unknown in real-world applications. Another method is to assign for each hypothesis a factor decaying from the time it is created. When the factor decreases through a threshold η, the corresponding hypothesis is removed from $\boldsymbol{H}$, which is pretty much similar to a queue with dynamic capacity. The challenge raised by this method is how to determine η, which could hardly be determined by cross-validation in an online learning scenario.

7.4.4 Complexity

Time and space complexity are of critical importance for designing real-time online learning algorithms. It is expected that an algorithm learns from data streams as quickly as possible such that it can keep pace with the data stream that could be of very high speed. It is also desirable that the algorithm does not occupy significantly large memory because of the concern of scalability. From a view of slightly high level, time and space complexity of REA should be related to (i) the difference between post-balanced ratio f and imbalanced ratio γ, that is, $f - \gamma$; (ii) k parameter of k-nearest neighbor; and (iii) the capacity of hypothesis set $\boldsymbol{H}$, that is, $|\boldsymbol{H}|$.

To quantify the time and space complexity of algorithms introduced in Section 7.3, their time and space consumption for learning from SEA (40 chunks), ELEC (40 chunks), and SHP (100 chunks) are recorded as in Tables 7.3–7.5, respectively. The hardware configuration is Intel Core i5 Processor with 4 GB random access memory (RAM).

Table 7.3 Time and Space Complexity for SEA Dataset

Complexity	Normal	SMOTE	UB	SERA	REA
Training time complexity (s)	2.632	4.092	7.668	3.952	5.356
Testing time complexity (s)	0.088	0.148	0.236	0.156	1.904
Space complexity (kb)	1142	1185	1362	1266	1633

Table 7.4 Time and Space Complexity for ELEC Dataset

Complexity	Normal	SMOTE	UB	SERA	REA
Training time complexity (s)	1.376	1.712	2.844	1.824	2.028
Testing time complexity (s)	0.084	0.148	0.244	0.14	1.744
Space complexity (kb)	129	142	155	151	248

Table 7.5 Time and Space Complexity for SHP Dataset

Complexity	Normal	SMOTE	UB	SERA	REA
Training time complexity (s)	10.29	17.89	60.10	16.98	21.62
Testing time complexity (s)	0.23	0.46	0.53	0.43	11.58
Space complexity (kb)	6881	7061	7329	9858	10,274

Considering the empirical performance of all algorithms on various datasets, it could be concluded that SERA has the best performance–complexity trade-off. That being said, as the consumption of time and space of REA is not significantly larger than that of SERA, applications that do not place strong requirements on speed and limitations on computational resources should as well consider using REA because of its exceptional learning performance.

7.5 CONCLUSION

This chapter reviews algorithms for learning from nonstationary data streams with imbalanced class distribution. All these algorithms basically manage to increase the number of minority class data within the data chunk under consideration to compensate the imbalanced class ratio, which can be categorized into the ones using an over-sampling technique to create synthetic minority class instances, that is, SMOTE/Learn++ and the others using previous minority class examples UB/SERA/MuSeRA/REA. The second category can be further divided into the ones using *all* previous minority class examples and the others using *partial* previous minority class examples. The key idea of using partial previous minority class examples to compensate imbalanced class ratio is to find a measurement to calculate the similarity degree between each previous minority class example and the current minority class set, which splits the algorithms in this subcategory into the ones using Mahalanobis distance and the others using k-nearest neighbor. This finally ends the taxonomy of algorithms.

Algorithms are introduced by referring to their algorithmic pseudo-codes, followed by the theoretical study and simulations on them. The results show that REA seems to be able to provide the most competitive results over other algorithms under comparison. Nevertheless, considering the efficiency in practice for stream data-based applications, SERA may provide the best trade-off between performance and algorithm complexity.

There are many interesting directions that can be followed up to pursue the study of learning from nonstationary stream with imbalanced class distribution. First of all, for REA/MuSeRA algorithms, an efficient and concrete mechanism is desired to enable them to remove the hypotheses with obsoleted knowledge on the fly to account for limited resources availability as well as concept drifts. For instance, one can explore integrating the method using Learn++ to prune the hypothesis into REA/MuSeRA. Second, the issue of compensating the imbalanced class ratio can be worked against the other way around. In other words, one can choose to remove less important majority class examples instead of explicitly increasing the minority class data. The effect would be the same, and the benefit is obvious, which is the avoidance of accommodating synthetic/previous data into the data chunk to potentially impair the integrity of target concept. The random under-sampling method employed by UB could be considered as a kind of preliminary effort to implement this. Finally, there seems to be no record indicating the usage of the cost-sensitive learning framework to address the problem. One could directly assign different misclassification costs to minority class and majority class examples during training in the hope for better learning performance. Aside from this naive implementation, a smarter way is to assign different misclassification costs to minority class examples, majority class examples, *and previous minority class examples*. The misclassification costs for previous minority class examples can even be set nonuniformly according to how similar they are with the minority class set in the training data chunk under consideration.

7.6 ACKNOWLEDGMENTS

This work was supported in part by the National Science Foundation (NSF) under grant ECCS 1053717 and Army Research Office (ARO) under Grant W911NF-12-1-0378.

REFERENCES

1. B. Babcock, S. Badu, M. Datar, R. Motwani, and J. Wisdom, "Models and issues in data stream systems," in *Proceedings of PODS*, (Madison, WI), ACM, 2002.

2. M. M. Gaber, S. Krishnaswamy, and A. Zaslavsky, "Adaptive mining techniques for data streams using algorithm output granularity," in *Workshop (AusDM 2003), Held in conjunction with the 2003 Congress on Evolutionary Computation (CEC 2003)*, (Newport Beach, CA), Springer-Verlag, 2003.

3. C. Aggarwal, *Data Streams: Models and Algorithms*. Berlin: Springer Press, 2007.

4. A. Sharma, "A note on batch and incremental learnability," *Journal of Computer and System Sciences*, vol. 56, no. 3, pp. 272–276, 1998.

5. S. Lange and G. Grieser, "On the power of incremental learning," *Theoretical Computer Science*, vol. 288, no. 2, pp. 277–307, 2002.

6. M. D. Muhlbaier, A. Topalis, and R. Polikar, "Learn++.nc: Combining ensemble of classifiers with dynamically weighted consult-and-vote for efficient incremental learning of new classes," *IEEE Transactions on Neural Networks*, vol. 20, no. 1, pp. 152–168, 2009.

7. P. Domingos and G. Hulten, "Mining high-speed data streams," in *Proceedings of the International Conference KDD*, (Boston, MA), pp. 71–80, ACM Press, 2000.

8. R. Polikar, L. Udpa, S. Udpa, and V. Honavar, "Learn++: An incremental learning algorithm for supervised neural networks," *IEEE Transactions on System, Man and Cybernetics (C), Special Issue on Knowledge Management*, vol. 31, pp. 497–508, 2001.

9. H. He and S. Chen, "IMORL: Incremental multiple-object recognition and localization," *IEEE Transactions on Neural Networks*, vol. 19, no. 10, pp. 1727–1738, 2008.

10. Y. Freund and R. Schapire, "Decision-theoretic generalization of on-line learning and application to boosting," *Journal of Computer and Systems Sciences*, vol. 55, no. 1, pp. 119–139, 1997.

11. P. Angelov and X. Zhou, "Evolving fuzzy systems from data streams in real-time," in *IEEE Symposium on Evolving Fuzzy Systems* (Ambelside, Lake District, UK), pp. 29–35, IEEE Press, 2006.

12. O. Georgieva and D. Filev, "Gustafson–Kessel algorithm for evolving data stream clustering," in *Proc. Int. Conf. Computer Systems and Technologies for PhD Students in Computing* (Rousse, Bulgaria), ACM, Article 62, 2009.

13. D. Filev and O. Georgieva, "An extended version of the Gustafson–Kessel algorithm for evolving data stream clustering," in *Evolving Intelligent Systems: Methodology and Applications* (P. Angelov, D. Filev, and N. Kasabov, eds.), pp. 273–300, John Wiley and Sons, IEEE Press Series on Computational Intelligence, 2010.

14. S. Grossberg, "Nonlinear neural networks: Principles, mechanisms, and architectures," *Neural Networks*, vol. 1, no. 1, pp. 17–161, 1988.

15. W. N. Street and Y. Kim, "A streaming ensemble algorithm (SEA) for large-scale classification," in *Proceedings of the Seventh ACM SIGKDD International Conference Knowledge Discovery and Data Mining* (San Francisco, CA, USA), pp. 377–382, ACM Press, 2001.

16. W. Fan, "Systematic data selection to mine concept-drifting data streams," in *Proceedings of ACM SIGKDD International Conference Knowledge Discovery and Data Mining*, (Seattle, WA), pp. 128–137, ACM Press, 2004.

17. M. Last, "Online classification of nonstationary data streams," *Intelligent Data Analysis*, vol. 6, no. 2, pp. 129–147, 2002.

18. Y. Law and C. Zaniolo, "An adaptive nearest neighbor classification algorithm for data streams," in *Proceedings of European Conference PKDD* (Porto, Portugal), pp. 108–120, Springer-Verlag, 2005.

19. H. He and E. A. Garcia, "Learning from imbalanced data," *IEEE Transactions on Knowledge and Data Engineering*, vol. 21, no. 9, pp. 1263–1284, 2009.

20. W. Fan, S. J. Stolfo, J. Zhang, and P. K. Chan, "Adacost: Misclassification cost-sensitive boosting," in *Proceedings of 16th International Conference on Machine Learning* (Bled, Slovenia), pp. 97–105, Morgan Kaufmann, 1999.

21. N. V. Chawla, K. W. Bowyer, L. O. Hall, and W. P. Kegelmeyer, "SMOTE: Synthetic minority over-sampling technique," *Journal of Artificial Intelligence Research*, vol. 16, pp. 321–357, 2002.

22. N. V. Chawla, A. Lazarevic, L. O. Hall, and K. W. Bowyer, "Smoteboost: Improving prediction of the minority class in boosting," in *Proceedings of the Principles of Knowledge Discovery in Databases, PKDD-2003*, (Cavtat-Dubrovnik, Croatia), pp. 107–119, Springer Press, 2003.

23. X. Hong, S. Chen, and C. J. Harris, "A kernel-based two-class classifier for imbalanced data-sets," *IEEE Transactions on Neural Networks*, vol. 18, no. 1, pp. 28–41, 2007.

24. Masnadi-Shirazi and N. Vasconcelos, "Asymmetric boosting," in *Proceedings of International Conference Machine Learning*, (Corvallis, OR), pp. 609–619, ACM, 2007.

25. R. P. G. Ditzler and N. Chawla, "An incremental learning algorithm for nonstationary environments and class imbalance," in *International Conference on Pattern Recognition (Istanbul, Turkey)* (P. Angelov, D. Filev, and N. Kasabov, eds.), pp. 2997–3000, IEEE, 2010.

26. J. Gao, W. Fan, J. Han, and P. S. Yu, "A general framework for mining concept-drifting streams with skewed distribution," in *Proceedings of International Conference SIAM (Minneapolis, MN, USA)*, ACM, 2007.

27. S. Chen and H. He, "SERA: Selectively recursive approach towards nonstationary imbalanced stream data mining," *IEEE-INNS-ENNS International Joint Conference on Neural Networks*, vol. 0, pp. 522–529, 2009.

28. S. Chen and H. He, "MuSeRA: Multiple selectively recursive approach towards imbalanced stream data mining," in *Proceedings of World Conference Computational Intelligence* (Barcelona, Spain), IEEE, 2010.

29. S. Chen and H. He, "Towards incremental learning of nonstationary imbalanced data stream: A multiple selectively recursive approach," *Evolving Systems*, vol. 2, no. 1, pp. 35–50, 2011.

30. P. C. Mahalanobis, "On the generalized distance in statistics," in *Proceedings of National Institute of Science of India*, vol. 2, no. 1, pp. 49–55, 1936.

31. J. Gao, W. Fan, and J. Han, "On appropriate assumptions to mine data streams: Analysis and practice," in *Proceedings of International Conference Data Mining* (Washington, DC, USA), pp. 143–152, IEEE Computer Society, 2007.

32. H. Wang, W. Fan, P. S. Yu, and J. Han, "Mining concept-drifting data streams using ensemble classifiers," in *KDD '03: Proceedings of the Ninth ACM SIGKDD International Conference on Knowledge Discovery and Data Mining* (Washington, DC, USA), pp. 226–235, ACM, 2003.

33. L. Breiman, "Bootstrap aggregating," *Machine Learning*, vol. 24, no. 2, pp. 123–140, 1996.

34. K. Tumer and J. Ghosh, "Error correlation and error reduction in ensemble classifiers," *Connection Science*, vol. 8, no. 3–4, pp. 385–403, 1996.

35. K. Tumer and J. Ghosh, "Analysis of decision boundaries in linearly combined neural classifiers," *Pattern Recognition*, vol. 29, pp. 341–348, 1996.

36. T. Fawcett, "ROC graphs: Notes and practical considerations for data mining researchers," Tech. Rep. HPL-2003-4, HP Laboratories, 2003.

37. M. Harries, "Splice-2 comparative evaluation: Electricity pricing," Tech. Rep. UNSW-CSE-TR-9905, The University of South Wales, 1999.

8

ASSESSMENT METRICS FOR IMBALANCED LEARNING

NATHALIE JAPKOWICZ

School of Electrical Engineering and Computer Science, University of Ottawa, Ottawa, ON, Canada and Department of Computer Science, Northern Illinois University, Illinois, USA

Abstract: Assessing learning systems is a very important aspect of the data-mining process. As such, many different types of metrics have been proposed over the years. In most cases, these metrics were not designed with the class imbalance problem in mind. As a result, some of them turn out to be appropriate for this problem, while others are not. The purpose of this chapter is to survey existing evaluation metrics and discuss their application to class-imbalanced domains. We survey many of the well-known metrics that were not specifically designed to handle class imbalances as well as more recent metrics that do take this issue into consideration.

8.1 INTRODUCTION

Evaluating learning algorithms is not a trivial issue as it requires judicious choices of assessment metrics, error estimation methods, and statistical tests, as well as an understanding that the resulting evaluation can never be fully conclusive. This is due, in part, to the inherent bias of any evaluation tool and to the frequent violation of the assumptions they rely on [1]. In the case of class imbalances, the problem is even more acute because the default, relatively robust procedures used for unskewed data can break down miserably when the data is skewed.

Take, for example, the case of accuracy that measures the percentage of times a classifier predicts the correct outcome in a testing set. This simple measure is ineffective in the presence of class imbalances as demonstrated by the

Imbalanced Learning: Foundations, Algorithms, and Applications, First Edition.
Edited by Haibo He and Yunqian Ma.
© 2013 The Institute of Electrical and Electronics Engineers, Inc. Published 2013 by John Wiley & Sons, Inc.

extreme case where, say, 99.9% of all the data is negative and only 0.1% positive. In such a case, the rough method, which consists of predicting "negative" in all cases, would produce an excellent accuracy rate of 99.9%. Obviously, this is not representative of what the classifier is really doing because, as suggested by its 0% recall, it is not an effective classifier at all, specifically if what it is trying to achieve is the recognition of rare, but potentially important events.

The inappropriateness of standard approaches is also apparent when considering error-estimation approaches. A 10-fold cross-validation, in particular, the most commonly used error-estimation method in machine learning, can easily break down in the case of class imbalances, even if the skew is less extreme than the one previously considered. For example, if 80% of the data is negative and 20% positive, when uniformly sampling the data in order to create the folds, it is quite likely that the distribution within each fold varies widely, with a possible dearth of positive data altogether in some extreme cases in which the dataset happens to be very small.

The behavior of statistical tests is altered as well by the presence of class imbalances, as the skew in the data can cause these tests to display different biases in different circumstances.

In this chapter, we will focus on the aspect of evaluation that concerns the choice of an assessment metric. This is, by far, the issue that has generated the greatest amount of attention in the community, as its answer is not as simple as the one pertaining to error estimation, where an easy solution consists of using stratified rather than purely random samplings of the data to create the folds (although there are other more complex issues with error estimation [1] and [2]); nor is it as complex as in the case of statistical testing, where the effect of class imbalances on a test can only be seen through numerous experiments and was not investigated at great length. We will, therefore, only touch upon the issues of error estimation and statistical testing briefly in the conclusion.

This chapter thus concentrates mainly on describing both metrics and graphical methods used in the case of class imbalances, concentrating on well-established methods and pointing out the newer experimental ones. Binary class problems will be the main focus, and the reader will be directed toward the literature that discusses extensions to multiple-class problems only in the conclusion.

The remainder of this chapter is organized as follows. In Section 8.2, we present an overview of the three families of assessment metrics used in machine learning—*threshold metrics, ranking methods and metrics* and *probabilistic metrics*—and discuss their general appropriateness to class imbalance situations. In Section 8.3, we focus on the threshold metrics particularly suited for imbalanced datasets. Section 8.4 looks at ranking methods and metrics often used in class-imbalanced situations. The probabilistic metrics did not seem specifically suited to the class imbalance problem; so we do not discuss them here. Finally, Section 8.5 concludes the chapter, looking at both other aspects of the classifier evaluation process that could be impacted on by class imbalances, and the case of multi-class-imbalanced problems.

8.2   A REVIEW OF EVALUATION METRIC FAMILIES AND THEIR APPLICABILITY TO THE CLASS IMBALANCE PROBLEM

Several machine learning researchers have identified three families of evaluation metrics used in the context of classification [3, 4]. These are the *threshold metrics* (e.g., accuracy and *F*-measure), the *ranking methods and metrics* [e.g., receiver operating characteristics (ROC) analysis and AUC), and the *probabilistic metrics* (e.g., root-mean-squared error). The purpose of this section is to discuss the advantages and disadvantages of these families with respect to the class imbalance problem.

In assessing both the families and the specific metrics, one must keep in mind the purpose of the evaluation process. For example, because Ferri et al. [3] considers that decreasing the overall value of an algorithm's worth because of its poor performance on one or a few infrequent classes is a nuisance, their conclusions are opposite to those typically reached by researchers who focus on problems with class imbalances or cost issues and for whom good performance on the majority class is usually of lesser consequence than good performance on the rarer or more important class. In this chapter, we are in agreement with the latter class of researchers, and so, in contrast to Ferri et al. [3], we take the position that sensitivity to the misclassification of infrequent classes is an asset rather than a liability.

As mentioned in [5], many studies have documented the weakness of the most notable threshold metric, accuracy, in comparison to ranking methods and metrics (most notably ROC analysis/AUC) in the case of class imbalances. The first such study, which in fact brought ROC analysis to the attention of machine learning researchers, is the one by Provost and Fawcett [6]. The main thrust of their article is that because the precise nature of the environment in which a machine learning system will be deployed is unknown, arbitrarily setting the conditions in which the system will be used is flawed. They argue that a typical measure such as accuracy does just that because it assumes equal error costs and constant and relatively balanced class priors, despite the fact that the actual conditions in which the system will operate are unknown. Their argument applies to more general cases than accuracy, but that is the metric they focus on in their discussion. Instead of systematically using the accuracy, they propose an assessment based on ROC analysis that does not make any assumptions about costs or priors, but rather, can evaluate the performance of different learning systems under all possible cost and prior situations. In fact, they propose a hybrid learning system based on the idea that suggests a different learner for each cost and prior condition.

A more recent study by Ferri et al. [3] expands this discussion to a large number of metrics and situations. It is the largest scale study, to date, which pits the different metric families and individual metrics against one another in various contexts (including the class imbalance one).[1] Their study is twofold. In

---

[1] A little earlier, Caruana and Niculescu-Mizil [4] also conducted a large-scale study of the different families of metrics, but they did not consider the case of class imbalances. We, thus, do not discuss their work here.

the first part, they determine the correlation between the different metrics and their family to other metrics and families, in various contexts; in the second part, they conduct a sensitivity analysis to study the specific performance of the group of metrics and their families that they studied in each context. We will focus only, herein, on the results they obtained in the context of the class imbalance problem.

The correlation study by Ferri et al. [3] shows that while the metrics are well correlated within each family in the balanced situation, this is not necessarily the case in the imbalanced situation. Indeed, two observations can be made: first, whatever correlations exist within a same family in the balanced case, these correlations are much weaker in the imbalanced situation, and second, the crossovers from one family to the next are different in the balanced and imbalanced cases. Without going into the details, these observations (and the first one in particular) allow us to conclude that the choice of metrics in the imbalanced case is of particular importance.

To analyze their sensitivity analysis results, we invert the conclusions of Ferri et al. [3], as previously mentioned, because we adopt the point of view that sensitivity to misclassification in infrequent classes is an asset rather than a liability. Note, however, that this inversion strategy should only be seen as a first approximation because the study by Ferri et al. [3] plots the probability of a wrong classification (as the class imbalance increases) no matter what class is considered and there is no way to differentiate between the different kinds of errors the classifiers make. Optimally, the study should be repeated from the perspective of researchers on the class imbalance problem. That being said, their study indicates that the rank-based measures behave best followed by some instances of threshold metrics. The probabilistic metrics that do behave acceptably well in the class imbalance case do so in the same spirit as some of the threshold metrics discussed later [the multi-class focused ones (Sections 8.3.5 and 8.3.6)]. In other words, it is not their probabilistic quality that makes them behave well in imbalanced cases, and therefore, they are not discussed any further in this chapter. The remainder of this chapter thus focuses only on the threshold metrics and ranking metrics most appropriate to the class imbalance problem. In the next two sections, we will discuss in great detail the measures that present specific interest for the class imbalance problem. We consider each class of metrics separately and discuss the particular metrics within these classes that are well suited to the class imbalance problem.

8.3   THRESHOLD METRICS: MULTIPLE- VERSUS SINGLE-CLASS FOCUS

As discussed in [1], threshold metrics can have a multiple- or a single-class focus. The multiple-class focus metrics consider the overall performance of the learning algorithm on all the classes in the dataset. These include accuracy, error rate, Cohen's kappa, and Fleiss' kappa measures. Precisely because of this multi-class focus, as also seen in the study by Ferri et al. [3], because the varying degree of importance on the different classes is not considered, performance metrics in this category do not fare very well in the class-imbalanced situation unless the class ratio is specifically taken into consideration. Single-class focus metrics, on the other hand, can be more sensitive to the issue of the varying degree of importance placed on the different classes and, as a result, be naturally better suited to evaluation in class-imbalanced domains. The single-class focus measures that are discussed in this section are: sensitivity/specificity, precision/recall, Geometric mean (G-mean), and F-measure. In addition to single-class focus metrics, we will discuss the multi-class focus metrics that take class ratios into consideration as a way to mitigate the contribution of the components on the overall results. We will also present a survey of more experimental metrics that were recently proposed but have not yet enjoyed much exposure in the community.

All the metrics discussed in this section are based on the concept of the confusion matrix. The confusion matrix for classifier f records the number of examples of each class that were correctly classified as belonging to that class by classifier f, as well as the number of examples of each class that were misclassified. For the misclassified examples, the confusion matrix considers all kinds of misclassification possible and records the number of examples that fall in each category. For example, if we consider a three-class problem, the following confusion matrix tells us that a examples of class A, e examples of class B, and i examples of class C were correctly classified by f. However, $b+c$ examples of class A were wrongly classified by f, b of which were mistakenly assigned to Class B, and c of which were mistakenly assigned to class C (and similarly for classes B and C).

	Predicted class A	Predicted class B	Predicted class C
Actual class A	a	b	c
Actual class B	d	e	f
Actual class C	g	h	i

In the binary class case, the above-mentioned matrix is reduced to a 2×2 format, and the issue of which class a misclassified example is assigned to disappears, as there remains only one possibility. In such a case, specific names are given to both the classes (positive and negative) and to the entries of the confusion matrix (true positive, false negative, false positive, and true negative) as shown in the following:

	Predicted positive	Predicted negative
Actual positive	True positive (TP)	False negative (FN)
Actual negative	False positive (FP)	True negative (TN)

The *true positive* and *true negative* entries indicate the number of examples correctly classified by classifier f as positive and negative, respectively. The *false negative* entry indicates the number of positive examples wrongly classified as negative. Conversely, the *false positive* entry indicates the number of negative examples wrongly classified as positive. With these quantities defined, we are now able to define the single-class focus metrics of interest here as well as the multiple-class focus metrics adapted to the class imbalance problem.

8.3.1 Sensitivity and Specificity

The *sensitivity* of a classifier f corresponds to its true positive rate or, in other words, the proportion of positive examples actually assigned as positive by f. The complement metric to this is called the *specificity* of classifier f and corresponds to the proportion of negative examples that are detected. It is the same quantity, only it is measured over the negative class. These two metrics are typically used in the medical context to assess the effectiveness of a clinical test in detecting a disease. They are defined as follows:

$$\text{Sensitivity} = \frac{\text{TP}}{\text{TP} + \text{FN}} \tag{8.1}$$

$$\text{Specificity} = \frac{\text{TN}}{\text{FP} + \text{TN}} \tag{8.2}$$

These measures, together, identify the proportions of the two classes correctly classified. However, unlike accuracy, they do this separately in the context of each individual class of instances. As a result, the class imbalance does not affect these measures. On the other hand, the cost of using these metrics appears in the form of a metric for each single class, which is more difficult to process than a single measure. Another issue that is missed by this pair of metrics is the measure of the proportion of examples assigned to a given class by classifier f that actually belongs to this class. This aspect is captured by the following pair of metrics.

8.3.2 Precision and Recall

The *precision* of a classifier f measures how precise f is when identifying the examples of a given class. More precisely, it assesses the proportion of examples assigned a positive classification that are truly positive. This quantity together with sensitivity, which is commonly called *recall* when considered together with precision, is typically used in the information retrieval context where researchers are interested in the proportion of relevant information identified along with the amount of actually relevant information from the information assessed as relevant by f. Precision and recall are defined as follows:

$$\text{Precision} = \frac{\text{TP}}{\text{TP} + \text{FP}} \qquad (8.3)$$

$$\text{Recall} = \frac{\text{TP}}{\text{TP} + \text{FN}} \qquad (8.4)$$

Here again, the focus is on the positive class only, meaning that the problems encountered by multi-class focus metrics in the case of the class imbalance problem are, once more, avoided. As for sensitivity and specificity, however, the cost of using precision and recall is that two measures must be considered and that absolutely no information is given on the performance of f on the negative class. This information did appear in the form of specificity in the previous pair of metrics.

We now turn our attention to a couple of ways of combining the components of the two pairs of single-focus metrics just discussed so as to obtain a single measure instead of a pair of metrics.

8.3.3 Geometric Mean

The G-mean was introduced by Kubat et al. [7] specifically as a response to the class imbalance problem and as a response to the fact that a single metric is easier to manipulate than a pair of metrics. This measure takes into account the relative balance of the classifier's performance on both the positive and the negative classes. In order to do so, it is defined as a function of both the sensitivity and the specificity of the classifier. G-mean is defined in more detail as follows:

$$G - \text{mean}_1 = \sqrt{\text{sensitivity} \times \text{specificity}} \qquad (8.5)$$

Because the two classes are given equal importance in this formulation, the G-mean, while more sensitive to class imbalances than accuracy, remains close, in some sense, to the multi-class focus category of metrics. Another version of the G-mean was, therefore, also suggested, which focuses solely on the positive class. In order to do so, it replaces the specificity term by the precision term, yielding

$$G - \text{mean}_2 = \sqrt{\text{sensitivity} \times \text{precision}} \qquad (8.6)$$

8.3.4 F-Measure

The F-measure is another combination metric whose purpose this time is to combine the values of the precision and recall of a classifier f on a given domain to a single scalar. It does so in a more sophisticated way than the G-mean, as it allows the user to weigh the contribution of each component as desired. More specifically, the F-measure is the weighted harmonic mean of precision and recall, and is formally defined as follows: For any $\alpha \in R$, $\alpha > 0$,

$$F_\alpha = \frac{(1 + \alpha)[\text{precision} \times \text{recall}]}{[\alpha \times \text{precision}] + \text{recall}} \qquad (8.7)$$

where α typically takes the values of 1, 2, or 0.5 to signify that precision and recall are equal in the case of $\alpha = 1$, that recall weighs twice as much as precision when $\alpha = 2$ and that precision weighs twice as much as recall when $\alpha = 0.5$. The F-measure is very popular in the domains of information retrieval or text categorization.

We now get back to multi-class focus metrics but show how their shortcomings in the case of class imbalances can be overcome with appropriate weightings.

8.3.5 Macro-Averaged Accuracy

The macro-averaged accuracy (MAA) is presented in [3] and is calculated as the *arithmetic* average of the partial accuracies of each class. Its formula is given as follows for the binary case ([3] gives a more general definition for multi-class problems):

$$\text{MAA} = \frac{\text{sensitivity} + \text{specificity}}{2} \qquad (8.8)$$

In this formulation, we see that no matter what the frequency of each class is, the classifier's accuracy on each class is equally weighted.

The second macro-averaged accuracy presented in [3] is the macro-averaged accuracy calculated as the *geometric* average of the partial accuracies of each class and is nothing but the G-mean that was already presented in Section 8.3.3. Both measures were shown to perform better than other threshold metrics in class-imbalanced cases in the experiments from [3] that we related in Section 8.2.

8.3.6 Newer Combinations of Threshold Metrics

Up to this point, we have only discussed the most common threshold metrics that deal with the class imbalance problem. In the past few years, a number of newer combinations of threshold metrics have been suggested to improve on the ones previously used in the community. We discuss them now.

8.3.6.1 Mean-Class-Weighted Accuracy
The mean-class-weighted accuracy [8] implements a small modification to the macro-average accuracy metric just presented. In particular, rather than giving sensitivity and specificity equal weights, Cohen et al. [8] choose to give the user control over each component's weight. More specifically, Cohen et al. [8] who applied machine learning

methods to medical problems found that both the G-mean and the F-measure were insufficient for their purposes. On the one hand, they found that the G-mean does not provide a mechanism for giving higher priority to rare instances coming from the minority class; on the other hand, they found that while the F-measure allows for some kind of prioritization, it does not take into account the performance on the negative class. In medical diagnosis tasks, Cohen et al. [8] claim that what would be useful is a relative weighting of sensitivity and specificity. They are able to achieve this by using the macro-average, but adding a user-defined weight to increase the emphasis given to one class over the other. The measure they propose is called the *mean-class-weighted accuracy* (MCWA) and is defined as follows for the binary case:

$$\text{MCWA} = w \times \text{sensitivity} + (1 - w) \times \text{specificity} \qquad (8.9)$$

where w is a value between 0 and 1, which represents the weight assigned to the positive class. As for the macro-averages described in Section 8.3.5, the formula is also defined for multiple classes. This newer measure, they found, fits their purpose better than the other existing combinations.

In the next section, we present yet another variation on the sensitivity/specificity combination.

8.3.6.2 *Optimized Precision*

Optimized precision was proposed by Ranawana and Palade [9] as another combination of sensitivity and specificity. This time, the goal is not to weigh the contributions of the classes according to some domain requirement, but rather to optimize both sensitivity and specificity. The new metric is defined as follows:

$$\text{Optimized precision} = \text{specificity} \times N_\text{n} + \text{sensitivity}$$
$$\times N_\text{p} - \frac{|\text{specificity} - \text{sensitivity}|}{\text{specificity} + \text{sensitivity}} \qquad (8.10)$$

where N_n represents the number of negative examples in the dataset, and N_p represents the number of positive examples in the dataset. The authors show that this measure is preferable to precision and other single performance measures, both theoretically and in an experimental study, but they do not compare it to the other combination metrics that were already discussed in the previous sections.

8.3.6.3 *Adjusted Geometric Mean*

The adjusted geometric mean (AGm) was designed to deal with the type of class imbalance problems whose purpose is to increase the sensitivity while keeping the reduction of specificity to a minimum. This is in contrast with the problems about which we do not have to worry as much about a reduction in specificity. Batuwita and Palade [10], the designers of AGm, showed that commonly used measures in imbalanced datasets are not adequate for their problem of interest in the domain of bioinformatics. The G-mean may lead the researcher to choose a model that reduces specificity too much; as for the F-measure, although it could, theoretically, avoid this problem by manipulating its parameter, it is not clear what value that parameter should take when the misclassification costs are not exactly known. To overcome these problems, Batuwita and Palade [10] proposed the AGm. Its definition is as follows:

$$\text{AGm} = \frac{G - \text{mean} + \text{specificity} \times N_n}{1 + N_n}, \text{ if specificity} > 0 \qquad (8.11)$$

$$\text{AGm} = 0, \text{ if sensitivity} = 0 \qquad (8.12)$$

This metric is more sensitive to variations in specificity than in sensitivity. Furthermore, this focus on specificity is related to the number of negative examples in the dataset so that the higher the degree of imbalance, the higher the reaction to changes in specificity. The authors showed the advantages of this new measure as compared to the G-mean and the F-measure.

8.3.6.4 *Index of Balanced Accuracy*   The designers of the index of balanced accuracy (IBA) [11] respond to the same type of problem of the G-mean (as well as of the AUC) described by Batuwita and Palade [10]. In particular, they show that different combinations of sensitivity and specificity can lead to the same values for the G-mean. Their generalized IBA is defined as follows:

$$\text{IBA}_\alpha(M) = (1 + \alpha \times \text{Dom}) \times M \qquad (8.13)$$

where dominance (Dom) is defined as:

$$\text{sensitivity} - \text{specificity} \qquad (8.14)$$

where M is any metric, and α is a weighting factor designed to reduce the influence of the dominance on the result of a particular metric M. IBA_α was tested both experimentally on artificial and University of California, Irvine (UCI) data and theoretically. It is shown to have low correlation with accuracy (which is not appropriate for class-imbalanced datasets), but high correlation with G-mean and AUC (which are appropriate for class-imbalanced datasets). It is, therefore, indeed, geared at the right type of problems.

8.4 RANKING METHODS AND METRICS: TAKING UNCERTAINTY INTO CONSIDERATION

An important disadvantage of all the threshold metrics discussed in the previous section is that they assume full knowledge of the conditions under which the classifier will be deployed. In particular, they assume that the class imbalance present in the training set is the one that will be encountered throughout the operating life of the classifier. If that is truly the case, then the previously discussed metrics are appropriate; however, it has been suggested that information related to skew (as well as cost and other prior probabilities) of the data is generally not known. In such cases, it is more useful to use evaluation methods that enable visualization or summarization performance over the full operating range of the classifier. In particular, such methods allow for the assessment of a classifier's performance over all possible imbalance or cost ratios.

In this section, we will discuss ROC analysis, cost curves, precision–recall (PR) curves, as well as a summary scalar metric known as *AUC or AUROC* (Area under the ROC curve). In addition, newer and more experimental methods and metrics will be presented.

8.4.1 ROC Analysis

In the context of the class imbalance problem, the concept of ROC analysis can be interpreted as follows. Imagine that instead of training a classifier f only at a given class imbalance level, that classifier is trained at all possible imbalance levels. For each of these levels, two measurements are taken as a pair, the true positive rate (or sensitivity), whose definition was given in Equation 8.1, and the false positive rate (FPR) (or false alarm rate), whose definition is given by:

$$\text{FPR} = \frac{\text{FP}}{\text{FP} + \text{TN}} \tag{8.15}$$

Many situations may yield the same measurement pairs, but that does not matter as duplicates are ignored. Once all the measurements have been made, the points represented by all the obtained pairs are plotted in what is called the *ROC space*, a graph that plots the true positive rate as a function of the false positive rate. The points are then joined in a smooth curve, which represents the ROC curve for that classifier. Figure 8.1 shows two ROC curves representing the performance of two classifiers $f1$ and $f2$ across all possible operating ranges.[2]

The closer a curve representing a classifier f is from the top left corner of the ROC space (small false positive rate, large true positive rate) the better the performance of that classifier. For example, $f1$ performs better than $f2$ in the graph of Figure 8.1. However, the ideal situation of Figure 8.1 rarely occurs in practice. More often than not, one is faced with a situation as that of Figure 8.2, where one classifier dominates the other in some parts of the ROC space, but not in others.

The reason why ROC analysis is well suited to the study of class imbalance domains is twofold. First, as in the case of the single-class focus metrics of the previous section, rather than being combined together into a single multi-class focus metric, performance on each class is decomposed into two distinct measures. Second, as discussed at the beginning of the section, the imbalance

---

[2]Note that this description of ROC analysis is more conceptual than practical. The actual construction of ROC curves uses a single training set and modifies the classifier's threshold to generate all the points it uses to build the ROC curve. More details can be found in [1].

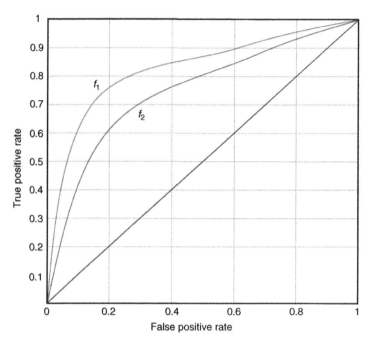

Figure 8.1   The ROC curves for two hypothetical scoring classifiers f_1 and f_2.

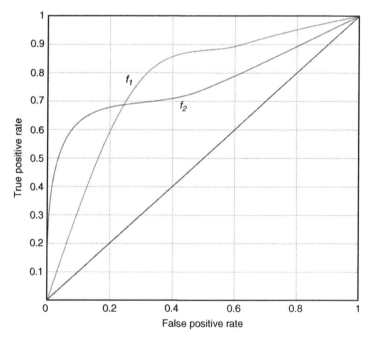

Figure 8.2   The ROC curves for two hypothetical scoring classifiers f_1 and f_2, in which a single classifier is not strictly dominant throughout the operating range.

ratio that truly applies in a domain is rarely precisely known. ROC analysis gives an evaluation of what may happen in these diverse situations.

ROC analysis was criticized by Webb and Ting [12] because of the problems that may arise in the case of changing distributions. The issue is that the true and false positive rates should remain invariant to changes in class distributions in order for ROC analysis to be valid. Yet, Webb and Ting [12] argue that changes in class distributions often also lead to changes in the true and false positive rates, thus invalidating the results. A response to Webb and Ting [12] was written by Fawcett and Flach [13], which alleviates their worries. Fawcett and Flach [13] argue that within the two important existing classes of domains, the problems pointed out by Webb and Ting [12] only apply to one of these classes, and then not always. When dealing with class imbalances in the case of changing distributions, it is thus recommended to be aware of the issues discussed in these two papers, before fully trusting the results of ROC analysis. In fact, Landgrebe et al. [14] point out another aspect of this discussion. They suggest that in imprecise environments, when class distributions change, the purpose of evaluation is to observe variability in performance, as the distribution changes in order to select the best model within an expected range of priors. They argue that ROC analysis is not appropriate for this exercise, and, instead, suggest the use of PR curves. PR curves are discussed in Section 8.4.3, which is followed in Section 8.4.5.2 by Landgrebe et al. [14]'s definition of PR curves summaries, which is somewhat similar to that of the AUC.

In the meantime, the next section addresses another common complaint about ROC analyses, namely that they are not practical to read in the case where a class imbalance or cost ratio is known.

8.4.2 Cost Curves

The issue of ROC analysis readability is taken into account by cost curves. What makes cost curves attractive is their ease of use in determining the best classifier to use in situations where the error costs or class distribution, or more generally the skew, are known. For example, in Figure 8.2, while it is clear that the curve corresponding to classifier f_2 dominates the curve corresponding to classifier f_1 at first and that the situation is inverted afterward, this information cannot easily be translated into information telling us for what costs and class distributions classifier f_2 performs better than classifier f_1. Cost curves do provide this kind of information.

In particular, cost curves plot the relative expected misclassification costs as a function of the proportion of positive instances in the dataset. An illustration of a cost curve plot is given in Figure 8.3. The important thing to keep in mind is that there is a point-line duality between cost curves and ROC curves. Cost curves are point-line duals of the ROC curves. In ROC space, a discrete classifier is represented by a point. The points representing several classifiers (produced by manipulating the threshold of the base classifier) can be joined (and extrapolated) to produce a ROC curve. In cost space, each of the ROC points is represented by

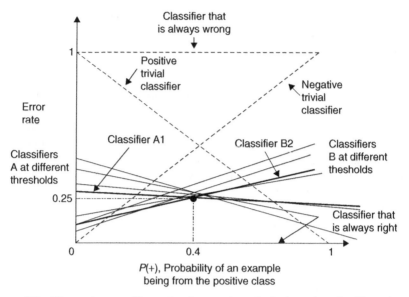

Figure 8.3 The cost curves illustration for two hypothetical scoring classifiers A_1 and B_2, in which a single classifier is not strictly dominant throughout the operating range.

a line, and the convex hull of the ROC space corresponds to the lower envelope created by all the classifier lines.

In cost space, the relative expected misclassification cost is plotted as a function of the probability of an example belonging to the positive class. Since the point in the ROC space where two curves cross is also represented in cost space by the crossing of two lines, this point can easily be found in cost space and its x-coordinate read off the graph. This value corresponds to the imbalance ratio at which a switch from the classifier that dominated the ROC space up to the curve crossing (say, classifier A) to the classifier that started dominating the ROC space afterward (say, classifier B) is warranted. The cost curves thus make it very easy to obtain this information. In contrast, ROC graphs tell us that, sometimes, A is preferable to B, but one cannot read off when this is so, from the ROC graph.

8.4.3 Precision–Recall Curves

PR curves are similar to ROC curves in that they explore the trade-off between the well-classified positive examples and the number of misclassified negative examples. As the name suggests, PR curves plot the precision of the classifier as a function of its recall, as shown in Figure 8.4. In other words, it measures the amount of precision that can be obtained as various degrees of recall are considered. For instance, in the domain of document retrieval systems, PR curves would plot the percentage of relevant documents identified as relevant against the percentage of relevant documents deemed as such with respect to all the

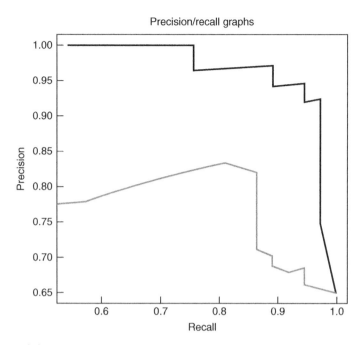

Figure 8.4   The precision–recall curves illustration for two scoring classifiers.

documents in the sample. The curves, thus, look different from ROC curves as they have a negative slope. This is because precision decreases as recall increases. PR curves are a popular visualization technique in the information retrieval field as illustrated by our earlier examples that discussed the notions of precision and recall. Further, it has been suggested that PR curves are sometimes more appropriate than ROC curves in the event of highly imbalanced data [15].

8.4.4   AUC

The AUC represents the performance of a classifier averaged over all the possible cost ratios. Noting that the ROC space is a unit square, it can be clearly seen that the AUC for a classifier f is such that $AUC(f) \in [0, 1]$ with the upper bound attained for a perfect classifier (one with TPR = 1 and FPR = 0). Moreover, as can be noted, the random classifier represented by the diagonal cuts the ROC space in half and hence $AUC(f_{random}) = 0.5$. For a classifier with a reasonably better performance than random guessing, we would expect to have an AUC greater than 0.5.

Elaborate methods have been suggested to calculate the AUC. However, using the Wilcoxon's rank sum statistic, we can obtain a simpler manner of estimating the AUC for ranking classifiers. To the scores assigned by the classifier to each test instance, we associate a rank in the order of decreasing scores. That is, the example with the highest score is assigned the rank 1. Then, we can calculate the AUC as:

$$\text{AUC}(f) = \frac{\sum_{i=1}^{|T_p|}(R_i - i)}{|T_p||T_n|}$$

where $T_p \subset T$ and $T_n \subset T$ are, respectively, the subsets of positive and negative examples in test set T, and R_i is the rank of the ith example in T_p given by classifier f.

AUC basically measures the probability of the classifier assigning a higher rank to a randomly chosen positive example than a randomly chosen negative example. Even though the AUC attempts to be a summary statistic, just as other single metric performance measures, it too loses significant information about the behavior of the learning algorithm over the entire operating range (for instance, it misses information on concavities in the performance, or trade-off behaviors between the TP and FP performances).

It can be argued that the AUC is a good way to get a score for the general performance of a classifier and to compare it to that of another classifier. This is particularly true in the case of imbalanced data where, as discussed earlier, accuracy is too strongly biased toward the dominant class. However, some criticisms have also appeared warning against the use of AUC across classifiers for comparative purposes. One of the most obvious ones is that if the ROC curves of the two classifiers intersect (such as in the case of Figure 8.2), then the AUC-based comparison between the classifiers can be relatively uninformative and even misleading. However, a possibly more serious limitation of the AUC for comparative purposes lies in the fact that the misclassification cost distributions (and hence the skew-ratio distributions) used by the AUC are different for different classifiers. This is discussed in the next subsection, which generally looks at newer and more experimental ranking metrics and graphical methods.

8.4.5 Newer Ranking Metrics and Graphical Methods

8.4.5.1 The H-measure The more serious criticism of AUC just mentioned means that, when comparing different classifiers using the AUC, one may in fact be comparing oranges and apples, as the AUC may give more weight to misclassifying a point by classifier A than it does by classifier B. This is because the AUC uses an implicit weight function that varies from classifier to classifier. This criticism was made by Hand [16], who also proposed the H-measure to remedy this problem. The H-measure allows the user to select a cost-weight function that is equal for all the classifiers under comparison and thus allows for fairer comparisons. The formulation of the H-measure is a little involved and will not be discussed here. The reader is referred to [16] for further details about the H-measure as well as a pointer to R code implementing it.

It is worth noting, however, that Hand [16]'s criticism was recently challenged by Flach et al. [17], who found that the criticism may only hold when the AUC is interpreted as a classification performance measure rather than as a pure ranking measure. Furthermore, they found that under that interpretation, the problem appears only if the thresholds used to construct the ROC curve are assumed to be optimal. The model dependence of the AUC, they claim, comes only from these assumptions. They argue that in fact, Hand [16] did not need to make such assumptions, and they propose a new derivation that, rather than assuming optimal thresholds in the construction of the ROC curve, considers all the points in the dataset as thresholds. Under this assumption, Flach et al. [17] are able to demonstrate that the AUC measure is in fact coherent and that the H-measure is unnecessary, at least from a theoretical point of view.

An interesting additional result of Flach [17] is that the H-measure proposed by Hand [16] is, in fact, a linear transformation of the value that the area under the cost curve would produce. With this simple and intuitive interpretation, the H-measure would be interesting to pit against the AUC in practical experiments, although, given the close relation between the ROC Graph and cost curves, it may be that the two metrics yield very similar if not identical conclusions.

The next section suggests yet another metric that computes an Area under the Curve, but this time, the curve in question is a PR curve.

8.4.5.2   AU PREC and I AU PREC

8.4.5.2   AU PREC and I AU PREC   The measures discussed in this section are to PR curves what AUC is to ROC analysis or what the H-measure (or a close cousin of it) is to cost curves. Rather than a single one, two metrics are presented as a result of Landgrebe et al. [14]'s observations that in the case of PR curves, because precision depends on the priors, different curves are obtained for different priors. Landgrebe et al. [14], thus, propose two metrics to summarize the information contained in PR curves: one that computes the Area Under the PR curve for a single prior, the *AU PREC*, and one that computes the Area Under the PR curve for a whole range of priors, the *I AU PREC*. The practical experiments conducted in the paper demonstrate that the AU PREC and I AU PREC metrics are more sensitive to changes in class distribution than the AUC.

8.4.5.3   B-ROC   The final experimental method that we present in this chapter claims to be another improvement on ROC analysis. In their work on Bayesian-ROCs (B-ROCs), Cardenas and Baras [18] propose a graphical method that they claim is better than ROC analysis for analyzing classifier performance on heavily imbalanced datasets for three reasons. The first reason is that rather than using the true and false positive rates, they plot the true positive rate and the precision. This, they claim, allows the user to control for a low false positive rate, which is important in applications such as intrusion detection systems, better than ROC analysis. Such a choice, they claim, is also more intuitive than the variables used in ROC analysis. A second reason is similar to the one given by Landgrebe et al. [14] with respect to PR curves: B-ROC allows the plotting of different curves for different class distributions, something that ROC analysis does not allow. The third reason concerns the comparisons of two classifiers. With ROC curves, in order to compare two classifiers, both the class distribution and mis-classification costs must be known or estimated. They argue that misclassification costs are often difficult to estimate. B-ROC allows them to bypass the issue of misclassification cost estimation altogether.

8.5 CONCLUSION

The purpose of this chapter was to discuss the issue of classifier evaluation in the case of class-imbalanced datasets. The discussion focused on evaluation metrics or on graphical methods commonly used or more recently proposed and less known. In particular, it looked at the following well-known single-class focus metrics belonging to the threshold category of metrics: sensitivity and specificity, precision and recall, and their combinations: the G-mean and the F-measure, respectively. It then considered a multi-class focus threshold metric that combines the partial accuracies of both classes, the macro-average accuracy. In the threshold metric category, still, other new and more experimental combination metrics were surveyed: mean-class-weighted accuracy, optimized precision, the AGms, and the IBA. The chapter then discussed the following methods and metrics belonging to the ranking category of methods and metrics: ROC analysis, cost curves, PR curves, and AUC. Still in the ranking category, the chapter discussed newer and more experimental methods and metrics: the H-measure, Area under the Precision-Recall Curve and Integrated Area under the ROC Curve, and B-ROC analysis.

Mainly because there has been very little discussion on other aspects of classifier evaluation in the context of class imbalances than evaluation metrics and graphical methods, this chapter did not delve into the issues of error estimation and statistical tests, which are important aspects of the evaluation process as well. We will now say a few words about these issues that should warrant much greater attention in the future.

In the case of error estimation, the main method used in the machine learning community is 10-fold cross-validation. As mentioned in the introduction, it is clear that when such a method is used, because of the random partitioning of the data into 10 subsets, the results may be erroneous because some of the subsets may not contain many or even any instances of the minority class if the dataset is extremely imbalanced or very small. This is well documented, and it is common, in the case of class imbalances in particular, to use stratified 10-fold cross-validation, which ensures that the proportion of positive to negative examples found in the original distribution is respected in all the folds (See [1] for a description of this approach). Going beyond this simple issue, however, both Japkowicz and Shah [1] and Raeder et al. [2] discuss the issue of error estimation in the general case. They both conclude that while 10-fold cross-validation is quite reliable in most cases, more research is necessary in order to establish ideal re-sampling strategies at all times. Raeder et al. [2] specifically mention the class imbalance situation where they suggest that additional experiments are needed to be able to make the kind of statements they made in the general case, when class imbalances are present.

With respect to the issue of statistical testing, we found a single paper—the paper by Keller et al. [19]—that looked at the issue of class imbalances. In this paper, the authors show that a particular test, the bootstrap percentile test for the difference in F_1 measure between two classifiers, is conservative in the absence of a skew, but becomes optimistically biased for close models in the presence of a skew. This points out a potential problem with statistical tests in the case of imbalanced data, but it is too succinct a study for general conclusions to be drawn. Once again, much more research would be necessary on this issue to make stronger statements.

We conclude this chapter by mentioning one aspect of classifier evaluation that was not discussed so far: the issue of evaluation in imbalanced multi-class classification problems. There have been a number of extensions of some of the previously discussed methods for the case of multiple-class-imbalanced datasets. On the ranking method front, these include multi-class ROC graphs, which generate as many ROC curves as there are classes [20]; multi-class AUC, which computes the weighted average of all the AUCs produced by the Multi-class ROC graph just mentioned; and a skew-sensitive version of this Multi-class AUC [21]. On the threshold-metrics front, two metrics have been used in the community: misclassification costs and a multi-class extension of the G-mean (See references to these works in [5] and a description in [3].)

8.6 ACKNOWLEDGMENTS

The author gratefully acknowledges the financial support from the Natural Science and Engineering Council of Canada through the Discovery Grant Program.

REFERENCES

1. N. Japkowicz and M. Shah, *Evaluating Learning Algorithms: A Classification Perspective*. (New York, USA), Cambridge University Press, 2011.

2. T. Raeder, T. R. Hoens, and N. V. Chawla, "Consequences of variability in classifier performance estimates," in *ICDM* (Sydney, Australia), pp. 421–430, IEEE Computer Society, 2010.

3. C. Ferri, J. Haernandez-Orallo, and R. Modroiu, "An experimental comparison of performance measures for classification," *Pattern Recognition Letters*, vol. 30, pp. 27–38, 2009.

4. R. Caruana and A. Niculescu-Mizil, "Data mining in metric space: An empirical analysis of supervised learning performance criteria," in *Proceedings of the Tenth ACM SIGKDD International Conference on Knowledge Discovery and Data mining*, (Seattle, WA), pp. 69–78, 2004.

5. H. He and E. A. Garcia, "Learning from imbalanced data," *IEEE Transactions on Knowledge and Data Engineering*, vol. 21, no. 9, pp. 1263–1284, 2009.

6. F. J. Provost and T. Fawcett, "Robust classification for imprecise environments," *Machine Learning*, vol. 42, no. 3, pp. 203–231, 2001.

7. M. Kubat, R. C. Holte, and S. Matwin, "Machine learning for the detection of oil spills in satellite radar images," *Machine Learning*, vol. 30, pp. 195–215, 1998.

8. G. Cohen, M. Hilario, H. Sax, S. Hugonnet, and A. Geissbühler, "Learning from imbalanced data in surveillance of nosocomial infection," *Artificial Intelligence in Medicine*, vol. 37, no. 1, pp. 7–18, 2006.

9. R. Ranawana and V. Palade, "Optimized precision: A new measure for classifier performance evaluation," in *Proceedings of the IEEE Congress on Evolutionary Computation* (Vancouver, BC), pp. 2254–2261, IEEE Computer Society, 2006.

10. R. Batuwita and V. Palade, "A new performance measure for class imbalance learning. Application to bioinformatics problems," in *ICMLA* (Miami Beach, FL, USA), pp. 545–550, IEEE Computer Society, 2009.

11. V. García, R. A. Mollineda, and J. S. Sánchez, "Theoretical analysis of a performance measure for imbalanced data," in *ICPR* (Istanbul, Turkey), pp. 617–620, IEEE Computer Society, 2010.

12. G. I. Webb and K. M. Ting, "On the application of ROC analysis to predict classification performance under varying class distribution," *Machine Learning*, vol. 58, pp. 25–32, 2005.

13. T. Fawcett and P. A. Flach, "A response to Webb and Ting's on the application of ROC analysis to predict classification performance under varying class distribution'," *Machine Learning*, vol. 58, pp. 33–38, 2005.

14. T. Landgrebe, P. Paclik, and R. P. W. Duin, "Precision–recall operating characteristics (p-ROC) curves in imprecise environments," in *Proceedings of the Eighteenth International Conference on Pattern Recognition* (Hong Kong, China), pp. 123–127, IEEE Computer Society, 2006.

15. J. Davis and M. Goadrich, "The relationship between precision–recall and ROC curves," in *the Proceedings of the Twenty-Third International Conference on Machine Learning*, (Pittsburgh, Pennsylvania, USA), pp. 233–240, ACM, 2006.

16. D. J. Hand, "Measuring classifier performance: A coherent alternative to the area under the ROC curve," *Machine Learning*, vol. 77, no. 1, pp. 103–123, 2009.

17. P. Flach, J. Hernandez-Orallo, and C. Ferri, "A coherent interpretation of AUC as a measure of aggregated classification performance," in *Proceedings of the 28th International Conference on Machine Learning (ICML-11)* (New York, NY, USA), pp. 657–664, Omnipress, 2011.

18. A. Cardenas and J. Baras, "B-ROC curves for the assessment of classifiers over imbalanced data sets," in *Proceedings of the Twenty-First National Conference on Artificial Intelligence* (Boston, MA, USA), pp. 1581–1584, AAAI Press, 2006.

19. M. Keller, S. Bengio, and S. Y. Wong, "Benchmarking non-parametric statistical tests," in *NIPS*, (Vancouver, BC, Canada), pp. 464, MIT Press, 2005.

20. T. Fawcett, "An introduction to ROC analysis," *Pattern Recognition Letters*, vol. 27, pp. 861–874, 2006.

21. D. J. Hand and R. J. Till, "A simple generalisation of the area under the ROC curve for multiple class classification problems," *Machine Learning*, vol. 45, pp. 171–186, 2001.

INDEX

Imbalanced Learning: Foundations, Algorithms, and Applications, First Edition.
Edited by Haibo He and Yunqian Ma.
© 2013 The Institute of Electrical and Electronics Engineers, Inc. Published 2013 by John Wiley & Sons, Inc.

Printed and bound by CPI Group (UK) Ltd, Croydon, CR0 4YY

16/04/2025

14658363-0003